CROFTING YEARS

Francis Thompson

First Edition 1984
Reprinted 1986
Reprinted 1988
Reprinted 1989
Reprinted 1991
Reprinted 1993
Reprinted 1995
Revised Edition 1997

The Publisher acknowledges subsidy from the Scottish Arts Council towards the publication of this volume.

Cover Picture:
Interior of black house, Carbost, 1954
Courtesy of An Comunn Gaidhealach, Western Isles.

The Chapter end-pieces are Illuminations much enlarged, from The Book of Kells.

The paper used in this book is produced from renewable forests and is chlorine-free.

Printed and bound by Gwasg Dinefwr Press Ltd., Llandybie

CROFTING YEARS

Francis Thompson

Luath Press Ltd.
Edinburgh

BOOKS BY FRANCIS THOMPSON

Harris and Lewis - 1968
Harris Tweed - 1969
St. Kilda - 1970
The Uists and Barra - 1974
The Highlands and Islands - 1974
The Supernatural Highlands - 1976
Victoian & Edwardian Highlands in Old Photographs - 1976
Murder and Mystery in the Highlands - 1977
Portrait of the River Spey - 1979
The National Mod - 1979
Shell Guide to Northern Scotland - 1987
The Western Isles - 1988
Hebrides in Old Picture Postcards - 1989
In Hebrides Seas - 1994

Is treasa tuath na Tighearna

DEDICATION

To The People Of Bernera
&
The Women Of Braes.

What they began will one day be completed.

ACKNOWLEDGEMENTS

The author would like to express his appreciation of the interest taken by a number of people in this book. In particular, those who supplied photographs: George MacDonald, Stornoway, for permission to use the work of his late father, Angus M. MacDonald, Stornoway; Dr. Kenneth Robertson, erstwhile of Daliburgh, South Uist; Reg. Allen, of Castlebay, Barra; Angus MacArthur, Stornoway; Donald Morrison, Oban, and the Country Life Museum. Special thanks are due to Donnie MacLean, of *An Comunn Gaidhealach*, and John Murdo MacMillan, Stornoway, for their helpful comments.

FRANCIS THOMPSON was born and grew up in Stornoway, and now teaches at the Lews Castle College there. His interests have always been centred in the Highlands and Islands, but always in the context of Scotland as a nation with an identifiable and viable culture. His deeply held beliefs have involved him in Scottish national politics and the struggle to achieve full recognition of his beloved Gaelic language.

To that end, he was, in 1967, appointed Editor of the first bilingual newspaper published in Scotland for over half a century, and, in 1968, was a founder-director of *Club Leabhar* (The Highland Book Club).

Most of his many books have dealt with the Highlands and Islands, although he has also written on electrical installation, the subject which he teaches. In 1978 he received a Writer's Bursary from the Scottish Arts Council, which allowed him to produce both poetry and short stories.

Francis Thompson's interest in crofting has been lifelong. He has watched the changes in both crofting and crofters brought about by new regulations and laws, and is convinced that the crofting way of life still provides a very valuable resource for Scotland, and, perhaps, for the whole world.

CONTENTS

FOREWORD

One hundred years ago the pages of the British Press were filled with news of the disturbing events taking place in the Highlands and Islands of Scotland. Common folk whom Queen Victoria called 'my beloved Highlanders' were in direct opposition to the agents of British law, the police and soldiers of that same Queen. Confrontations were often bloody. Distressful scenes were enacted. Those who owned much of the land in the Highlands and Islands were seeing an old order being changed, to what, they could not forecast. Politicians saw in the unrest the seeds of revolution which could easily spread to the British working class. Thus it was that the eyes of those who controlled the huge British Empire were drawn to the north of Scotland.

And the reason? It was both simple and complex. Simple, in that the people who lived on the land owned by alienated clan chiefs and alien Victorian industrialists required a simple justice: security of tenure to provide a stable future for themselves and their children. It was complex in that the relationship of a landowner to his property, and the rights such possession implied, could both be eroded, and the way paved to further far-reaching changes in favour of the tenant.

This book is not so much concerned with the facts of history (which have been written about by sympathetic historians in recent years) but rather with the social background of those who were to become enshrined in British law as '*crofters*' in the Crofters' Act of 1886. The character of any community is derived largely from the daily life of its members; glimpses of that character will be found in the following pages. The centenary of the 1886 Act was recognised by a full assessment of the role of the crofter in the future and the environmental and economic function of crofting as a form of land use. The markers for the future laid down in 1986 are now being developed cautiously but with a determined articulation which will ensure that crofting retains its in-built strength to be the socio-economic sheet-anchor it has been for the people of the Highlands and Islands.

AN UNENVIABLE HISTORY

The Highlands and Islands of Scotland have never been strangers to the turbulent events of history. Seldom did a century pass without one major event occurring which disrupted the life of the folk of the Highlands, often painfully and with lasting effects. Perhaps no event was more destructive than the Jacobite Rising of 1745, which created such a turbulence that some of its waves lap on the shores of many Highland communities even today. The changes which the aftermath of that event introduced into the region affected both the landed clan chiefs and the landless commonalty, in particular that class of folk who were to be called '*crofters*'. They were to discover that their way of life stood at the threshold of an uncertain future, in which their destiny was to be carved by strangers and the decisions of politicians as far away as the Mother of Parliaments, in London.

> 'There is nothing in history so absolutely mean as the eviction of the Highlanders by chiefs solely indebted for every inch of land they ever held to the strong arms and trusty blades of the progenitors of those whom the effeminate and ungrateful chiefs of the nineteenth century have so ruthlessly oppressed, evicted and despoiled.'

So said Alexander MacKenzie, a prominent figure in the agitation to win security of tenure and legal recognition for the crofter population of the Highlands and Islands.

The phase in Highland history known as 'The Clearances' saw the removal of more than 500,000 people from the region. In Skye alone, for the period of four decades after 1840, more than 30,000 people were evicted from their holdings. Thousands were packed into unseaworthy vessels which never reached their destinations across the Atlantic, but found their merciful haven at the bottom of the sea. The process of forcibly removing the common folk from estates began in Sutherland, to spread like a disease to other areas.

People were, however, being evicted long before the factors in Sutherland went about their work. Dr. Samuel Johnson, visiting Skye in 1773, was moved to observe many instances of clearances in Skye. In the words of his companion James Boswell, '*When we reached the harbour of Portree there was laying in it a vessel to carry off the emigrants called the "Nestor". It made short settlement of the difference between a chief and his clan.*' Johnson said that '*a rapacious chief could make a wilderness of his estate.*' It was a time of change, when the control of the flood sluice-gates was in the hands of those to whom history had offered the chance for personal advantage.

Yet it was the folk who had offered even their lives in the service of King and Country who were eventually placed at a disadvantage. The Highlander, who had been foremost in securing and maintaining the supremacy of British interests in every quarter of the globe, was suddenly faced with an uncertain future at home. From 1740 to 1815, no less than fifty battalions of infantry had been raised mainly from the Highlands, besides many fencible or militia regiments. They fought in the Seven Years' War when Britain gained India and Canada for her growing Empire. They fought in the American War of Independence, and in the Peninsular and Napoleonic Wars that followed. They distinguished themselves with Wolfe at Quebec, with Sir John Moore at Corunna, and with Wellington at Waterloo. And it was the Highland Brigade, and in particular the 93rd Highlanders, the 'Thin Red Line', under the leadership of Sir Colin Campbell, which really saved the day at the Battle of Balaclava.

That record of service was, however, no guarantee of protection from those whom the Government now saw fit to recognise as legal proprietors of the land rather than as feudal superiors. This recognition changed the relationship between the chief and his people, a relationship which was already in the course of dilution, so that thousands of people found themselves at the mercy of landowners and their agents who required more and more money from their properties, at a time when hard cash was rare enough in the Highlands and Islands.

It was thus inevitable that conditions imposed on those who lived on the land, though they had no title to it in the context of codified law, would reach a flashpoint in a people who saw no future but what promised hardship, suffering, starvation and death. For those who had the means, emigration was the immediate answer, and they left the Highland shores in massive waves to provide the healthy and honest citizens of new countries across the Atlantic, folk who might well have provided the foundation of a new and vigorous Highland society had they been allowed to exercise their talents. For the rest, their only choice was to become submissive to the dictates of landlords and their factors. It has been observed that after the 1850's, which saw some of the worst clearances in the history of the Highlands and Islands, there followed some two decades of comparative quiet in the crofting areas, during which even the bravest of men fell silent in fear of eviction of themselves and their families.

'I am ashamed to confess it now that I trembled more before the factor than I did before the Lord of Lords.' So said Donald MacAskill at a Land Reform meeting in 1884. John Murdoch, of the 'Highlander' newspaper, visited the Gordon estate of South Uist in 1875 and found the crofters there in such a state of 'slavish fear' that, despite their many grievances, they would not complain to the factor in case he 'would drive them from the estate.'

Lacking a political solution to their problems, the people of the crofting areas turned to religion, and flocked to the new Free Church, formed after the Disruption of 1843, in such numbers that in most parish churches ministers of the established church found themselves preaching to empty pews. This move invited the inevitable reaction from the landlords, who denied all requests for sites for churches for the new-formed congregations. Lord MacDonald, the biggest landowner in Skye, consistently refused sites to the Snizort, Kilmuir and Portree congregations so that for many years their services were held in the open air. *'My predecessor often preached with the hailstones dancing on his forehead, his hearers wiping away the snow before they could sit down'*, the Rev. Joseph Lamont of Snizort once said.

In Strontian, Argyll, the congregation there solved the problem created by the landlord's continual refusal to grant them a site by using the famous 'floating church' moored in Loch Sunart. If anything, those years of travail, so characteristic of the formative decades of the Free Church and its adherents, only served to harden the attitude of the common tenantry against the landlords.

The haven of the new religion offered a training arena to those who were otherwise stifled by the landlords and their factors, a training arena in which they could get their breath back, regenerate their jaded spirits, and equip themselves with the skill of articulation, almost as though to prepare them for the war of words that was to become a significant element in the battle for crofting rights in the 1880's. Indeed, when the Napier Commission was set up in 1883 to examine the crofting problem, many of those who had become prominent in the Free Church as elders amazed the members of the Commission with their eloquence and forthrightness, and they carried their hearers with them by their conviction. However, only too many of these 'Men' (they were so called to distinguish them from the Ministers) devoted themselves to the furtherance of ecclesiastical affairs to such an extent that they eventually considered themselves above such secular matters as land reform and the setting to rights of crofting grievances. Thus it was that the leaders of the agitation in the Crofters' War were more to be found as exiled Highlanders from the large towns and cities than *bona fide* residents in the Highlands and Islands.

Eloquence was certainly needed, to speak out against the increasing volume of injustices being perpetrated against the crofting population.

The decades up to the 1880's were amongst the busiest for wealthy English and Scottish *nouveax riches* who vied with each other in the purchase of Highland estates, to play God over deer, salmon, grouse and human beings, albeit through the medium of their agents, the estate factors. It was the attitude and the actions of those factors which blew into life the smouldering fires of resentment against injustice. In Skye, to quote but one instance, many petty tyrannies were exercised by factors. The common method there of dealing with arrears of rent was to issue a summons for the removal of the defaulting tenant, instead of issuing a notice for the recovery of a small debt. And when a crofter was evicted, the incoming tenant was obliged to pay off his predecessor's arrears (purchasing the 'goodwill' of the croft was the term used to describe the technique). Even the collection of seaweed from the foreshore was often forbidden, or else had to be paid for. In the Broadford district of Skye, the right to seaware belonged to the tacksman, or main tenant land-holder, and crofters who could not afford to buy what they needed to fertilise their land saw their cropping returns grow less and less, and eventually faced starvation and pauperism.

Many petty restrictions were often placed on the crofters. In some areas they were not allowed to keep dogs which might put up game and deer, hunted by shooting tenants on the estates of absentee landlords. Deer were often forced down from the slopes to make a dash for freedom across the crofters' planted fields, with the inevitable damage to crops; this had to be endured without compensation. Another irritation was that deer grazing on grass and crops were also sacrosanct.

Other impositions included the right of the landlord or tacksman to demand free labour. Dr. MacGillivray, the owner of the island of Barra, demanded 60 days' labour for every acre of 'potato ground' held by cottars, who differed from crofters in that they possessed only a house of sorts and a small piece of ground for potatoes. In some areas the number of cottars exceeded the crofters. In 1886 in Skye, cottars numbered a quarter of the combined crofter/cottar population. Indeed, it could be said that the presence of cottar families in many areas added considerably to the problems of the crofters trying to make full use of their land, in that the sub-division of land on a croft to cater for their growing families assuming cottar status, took more land out of production, land that might have averted the problems of near-starvation which occurred from time to time. Occasionally the crofters tried to rectify the situation by requesting factors to take steps to remove the cottars, as occurred in Lewis in the early 1880's.

The seeds of revolt against their intolerable conditions were initially sowed in 1874, by crofters on the remote island of Great Bernera off the west coast of Lewis. Great Bernera was far removed from the hustle and bustle of Land Leaguers, newspapers, radicals and conservatives, and, indeed, far removed from the proliferation of Celtic and Gaelic-based societies which had begun to grow overnight like mushrooms.

The island of Lewis at that time was the property of Sir James Matheson, of the Far Eastern firm of Jardine, Matheson and Company, a man who had made a tidy fortune in opium-trafficking and whom Disraeli once called: '*....a dreadful man, richer than Croesus, one MacDrug, fresh from Canton with a million of opium in each pocket.*' Although he did spend some money on relief work during the famine years in Lewis, he also during those years erected for himself a castle at Stornoway, at a cost of £60,000. It was to this edifice that Lewis crofters were summoned to pay their rents and answer for their misdeeds. His factor was one Donald Munro, who had so many official hats that he could never remember all the posts he had acquired in the course of his administration of Lewis. Munro was little short of tyrannical in his dealings with the Lewis crofters. One crofter giving evidence before the Napier Commission stated that he was convinced that Munro's policy '*from the first day of his factorship to the last was to extirpate the people of Lewis so far as he could.*' If that was not quite Munro's intention, the statement indicated Munro's standing in the minds of the folk of Lewis. He had come to Lewis as a solicitor in 1842, and was appointed factor to Sir James

Matheson eleven years later; by 1874 he was involved in so great a number of offices that Professor John Stuart Blackie called him '*his polyonymous omnipotence*'.

Munro's Nemesis arrived on the Lewis scene in 1872, when the sporting estates of Morsgail and Scaliscro were formed by the taking over of the summer grazings of the Bernera crofters. These summer grazings were on the mainland of Lewis, across Loch Roag, for there were none suitable on Bernera itself. In compensation, the crofters were offered some of the moorland near the sea, which had formed part of the former farm of Earshader. Though the new grazings were not so good as the old lands, the crofters agreed to the exchange and also agreed to build a stone dyke as a barrier between the new grazings and the deer forest of Scaliscro. At a meeting a paper was read to the crofters, which the crofters signed, they being given to understand that the paper recorded the agreement. Most of the the crofters could only sign by making their marks. This paper was never produced at the subsequent trial of the Bernera crofters, and, indeed, Munro denied it had ever existed. Two years passed, during which the dyke, seven miles long, was built, and at the end of that time came the news that a new decision had been made. The crofters could no longer have the Earshader grazings, but they could have land elsewhere. Naturally the crofters resented this arbitrary treatment and they gave voice to their thoughts on the matter. Donald Munro himself went over to Bernera to talk to the crofters and, on seeing that little progress could be made, threatened to reinforce his decision by bringing the Volunteers from Stornoway (of which he was the Commanding Officer).

Munro also threatened to have the Bernera crofters evicted, even though none was in arrears of rent. Munro said later he had meant the threat about the Volunteers as a joke, but it was no joke when, on March 24th, 1874, estate officials came under orders from Munro with the intention of serving 58 summonses of removal. Their progress through the island of Bernera was made with no hindrance, until in the evening they were met by a crowd of boys and girls who threw grass divots at them. This treatment caused the sheriff officer to lose his temper and exclaim: '*If I had a gun with me, there would be some of the women in Bernera tonight lamenting their sons.*'

The following morning the party was confronted by a number of crofters who wanted to know the identity of the man who had threatened to shoot their children. A short struggle ensued during which the sheriff officer had his coat torn, which so incensed him that he lodged a complaint of assault against the crofters on his return to Stornoway. This led to the arrest of three of the Bernera crofters. Before their trial took place in Stornoway, other crofters from Bernera went to see Sir James Matheson about the affair. He expressed surprise at the intention of his factor to evict 58 crofters from the island. Munro said later that he did not think his intention to evict 58 crofters was a matter of sufficient importance to lay before his employer. In the event, the three arrested crofters were acquitted, and Munro's methods came in for severe criticism. The sheriff officer was himself fined 21 shillings for assault.

The so-called 'Bernera Riot' was a small enough affair, but it had important repercussions, for crofters in other areas of Lewis began to take their complaints about Munro straight to Sir James Matheson. Munro, in an effort to smooth things over, tried to extricate himself from the morass which had been a-growing for so many years. He even agreed to distribute to the folk of Lewis some £600 of the Ness Fund, which had been raised in 1863 when 31 men were drowned off the Butt of Lewis. The fund had been used by him for over ten years to pay the rents due by some of the widows who had succeeded to their husbands' crofts.

Thus exposed, Munro's positions were one by one taken from him until he was dismissed in 1875. In many ways he epitomised the Highland factor, the agent of an absentee landlord vested with the powers of a dictator and who used his position to his own advantage. Though it was to be another seven years before open resistance was to be made by crofters to their factors, the Bernera trial afforded the radical Press the opportunity to publicise the invidious position of crofters in the Highlands and Islands, and so create a sympathetic public opinion which was to be a considerable weapon of advantage a decade later when the cause of the crofting population eventually made its overdue appearance on the floors of both Houses of Parliament.

In October 1877 a peculiar event took place at the north end of Skye, on the property of Captain Fraser, owner of Uig Lodge. Two rivers, swollen by torrential rain, combined their courses and swept down to Uig Bay, destroying bridges, completely wrecking Uig Lodge (in which the estate manager was drowned) and so utterly devastating Uig cemetery that the bodies were found in the sea and in the gardens of the Lodge. As it happened, Captain Fraser was not known for his benevolent attitude towards the crofters on his estate, and the incident was dramatised by a correspondent of the newspaper 'Highlander', run by John Murdoch. No punches were pulled in an article which appeared in the columns of the paper:

> 'The belief is common throughout the parish that the disaster is a judgement upon Captain Fraser's property. It is very remarkable, it is said, that all the destruction on Skye should be on his estate. What looks so singular is that two rivers should break through every barrier and aim at Captain Fraser's house. Again, it is strange that nearly all the dead buried in Uig in the last five hundred years should be brought up as it were against his house, as if the dead in their graves rose to perform the work of vengeance which the living had not the spirit to execute. But although the living would not put forth a hand against the laird, they do not hesitate to express their regret that the proprietor was not in the place of the manager when he was swept away. It is sadder than the destruction itself that such feeling should be kindled under the land laws of Great Britain.'

This, of course, could not go unchallenged. Although Murdoch was not personally responsible for the appearance of the article, he being away at the time on a fund-raising mission, he took responsibility for the piece, and was faced with an action for damages of £1000. He got off with damages assessed at £50, with costs for legal expenses. The case was another victory on the progress towards the Crofters Act, and it served in another way: it was on Captain Fraser's lands that a dispute arose which was to spark off the 'troubles' in Skye in 1882, troubles at Valtos on the Kilmuir estate, and at the Braes, on Lord MacDonald's estate.

By 1882 the simmering pot of crofting troubles was ready for the final boil and it needed just one incident — perforce it had also to be violent — to make the contents spill over and create a situation where a decision, one way or the other, had to be taken to settle the issues involved. By coincidence the matter was brought to a head on an estate owned by Lord MacDonald, one of a decreasing number of traditional landowners whose forebears had once been able to call on their armed clansmen to fight for the clan lands and the honour of the MacDonalds. Far from being incomers, the MacDonalds had eight centuries of Highland history behind them, with many illustrious names in the family line. Yet, by the nineteenth century they commanded little loyalty from the crofters who often bore the same surname, for the old relationships had disappeared: Lord MacDonald was just another landlord.

His extensive estates were managed by a team of factors, one of whom, Alexander MacDonald, was a solicitor, and, like Donald Munro in Lewis, held a bewildering number of positions, ranging from being a member of numerous School Boards to being a captain in the local militia. He was also a landowner in his own right. It was to this man that some of the crofters from the MacDonald estate of Braes went to demand the restoration of the summer grazings on Ben Lee, which had been taken away from them in the 1860's. The demand was rejected out of hand and the crofters retaliated by declaring that they would withold payments of rents until the Ben Lee grazings were restored to them. The demand and its rejection were made in a fairly calm manner; its violent outcome was furthest from the minds of all concerned.

But the matter was not allowed to rest. Lord MacDonald, pondering the affair, decided that he and his position had been affronted, and so he went to law, demanding that his tenants be tried and convicted for criminal intimidation. In its wisdom the law decided there was no case to answer; but, true to its traditional form in the Highlands, advised Lord MacDonald to evict those whom he thought might be the ringleaders among the Braes tenants.

In April 1882 eviction notices were taken against a dozen or so men, to be delivered by a sheriff officer and two assistants. Before the trio came in sight of the Braes townships, they were met by over 100 people who took the summonses and compelled the sheriff officer to burn them, advising him not to return. In this way the final fuse in the long war between tenant and landlord was lit. As the sheriff officer had suffered deforcement, it presented the law with an ideal opportunity to '*stamp out quickly the first germs of anything like the Irish disease*'. In the legal mind there lurked the idea that if the Braes affair could be settled harshly it would serve as a once-for-all warning to any crofting community anywhere in the Highlands and Islands that the position of the landlord was unassailable.

It was thus decided to arrest named crofters from Braes, and to help him carry out the arrests the Sheriff of Inverness-shire, William Ivory, asked the Government for 100 soldiers. But the Government in its cautious wisdom, sent only 50 policemen from Glasgow north to Skye. With them came a battery of Press reporters, scenting blood in the outcome of the incident. The Press was there to support the cause of the landlord, but, unwittingly, as it turned out, the reporters provided the crofters with the best publicity machine possible.

The Press and Sheriff Ivory's police were on board the steamer as it approached Portree under cover of darkness. But Providence was on the side of the Braes crofters. The ship went aground in the first of a series of mishaps, so that the arresting party did not arrive at the Braes until it was daylight, and then went about its business of arresting without incident. It was the return journey that sparked off the powder keg.

On the return to Portree, the arresting party was met by a large crowd of people, amongst whom were women who '*with the most violent gestures and imprecations*' declared that the police must be attacked. Stones were thrown and the police retaliated by drawing batons and charging the crowd. *'Huge boulders darkened the horizon as they sped from the hands of infuriated men and women. Large sticks and flails were brandished and brought down with crushing force upon the police. Many were struck, and a number more or less injured.'* It was with the greatest difficulty that Sheriff Ivory's party managed to extricate themselves from the Braes crofters and reach the comparative safety of Portree. But even there they were hedged around by a crowd who contented themselves with shouting and hissing.

The trial of the Braes men took place in Inverness; they were found guilty and fined, but had their fines paid by supporters in court, and their legal costs paid by the Federation of Celtic Societies. They returned to Skye to see their fellow-crofters of Braes driving their stock onto the slopes of Ben Lee — and to see the cattle stay there despite writs issued from Edinburgh requiring the crofters to desist from appropriating Lord MacDonald's land in this way. They refused. And the law of Scotland was, for once, stopped in its tracks. It needed time to think, and when it had thought, its answer was the standard reflex action usually associated with British colonial administrations: Send for a gunboat.

The gunboat did arrive. But instead of the expected detachment of armed Marines, a Government representative was sent to require the company of some of the crofters from Glendale, where there had been a consistent refusal to pay rents and a persistent use of forbidden grazing land for their stock. The men of Glendale had also forced the police from the district in no uncertain manner. The Glendale 'Martyrs' were imprisoned in Edinburgh for two months. On their release in May of 1883, they made their way home to a new situation. The Government had at last yielded to pressure and had obtained, on 17 March, 1883, a Royal Warrant to establish the Napier Commission to investigate the whole of the crofting problem in the Highlands and Islands. It was the end of the beginning.

Before 1886, there was no Statutory legislation which was applicable to the crofting situation in Scotland. Even the term *'croft'* came slowly into being, and meant, in Gaelic, a small piece of arable land, with the *'crofter'* recognised as one who held a *'croft'* of land. The modern croft and crofter or small tenant seldom appears much before the beginning of the 18th century; they were tenants-at-will under the authority of the tacksmen and wadsetters, but in practice their tenure of the land they worked and lived on was reasonably secure: so long as a tenant was industrious and paid his rent in cash, in kind, or in work-service, he was allowed to occupy his house and land without fear of eviction. Seldom, in fact, did the actual proprietor of the land come into direct contact with the tenants, though records show that there was this close relationship between his people and Sir Donald MacDonald in Skye and in North Uist in 1718.

Prior to the Croftings Holdings (Scotland) Act of 1886, all crofts, farms and agricultural holdings in Scotland came under the ordinary agricultural law of Scotland. Under this law, a crofter was regarded as the tenant of a piece of land owned by a landlord; the tenant's occupation of the land was on a year-to-year basis and at a mutually agreed rent. This was the extent of the crofter's legal recognition, and it made no difference whether he and his forebears had lived in the same place for generations or was a new tenant of a year's standing. The law recognised the right of a landlord to terminate a crofter's tenancy at the end of any year after issuing 40 days' notice in writing. The landlord could then recover the land, together with any buildings or improvements thereon, and was under no legal obligation to offer money to the tenant for those improvements even though the latter had created them. The situation was of course different when the landlord had actually provided buildings and equipment for use by the tenant for an agreed price and for an agreed term of lease.

While this system worked reasonably successfully in other parts of Scotland, in the Highlands the working out of a terrible history was coming to a head, with the Clearances only one of a number of successive factors which threw into public awareness the complete inadequacy and the weakness of the law as it pertained to agricultural practice.

Much of the residual population had been shifted from their original holdings to make room for the more profitable sheep (and, later, deer), and they now clung — those who had not been cleared out of Scotland completely — to marginal lands next to the sea, on inferior soil, with imposed restrictions, such as, for example, forbidding the taking of seaware for use as fertiliser. The people had absolutely no security of tenure and lived in housing conditions which appalled even those to whom the urban slums of the Industrial Revolution were familiar. As the result of long and sore agitation for the improvement of the economic and social condition of the Highlander, the Napier Commission was appointed in March 1883. The Commission's remit was '*to enquire into the condition of the crofters and cottars in the Highlands and Islands of Scotland and all matters affecting the same or relating thereto.*' The Commissioners were authorised to examine witnesses, to require the production of documents and records and to make visits and inspections. In fact, the Commission travelled extensively throughout the crofting areas. The record of the many meetings held and all the oral evidence heard is contained in four hefty volumes, in addition to a fifth volume of over 500 pages of documentary evidence attached to the Commissioners' final report.

At first sight the members appointed to the Commission were rather unlikely men to investigate without bias the conditions in the Highlands, conditions which had been created by landed proprietors, of whom four were now Commissioners. Indeed, the welcome given to the Commission by Whig and Conservative newspapers boded the worst in the minds of those who had spent time and energy for over a decade in attempts to improve the lot of the crofters.

The Chairman was Lord Napier, a Border laird who, while he had no family connections with the Highlands, was fortunately a sympathetic observer of the Highland scene. Most of his life had been spent in the Diplomatic Service. After his retirement he became well known for his questions in the House of Lords on such issues as the amount of land in Scotland under cultivation, land used for deer forests and land capable of reclamation.

Sir Kenneth MacKenzie was the 6th Baronet of Gairloch, and had a reasonably good reputation as a benevolent landlord. He cleared off the arrears of rent on his estate by offering his tenants work in making new roads, on drainage schemes and on improving croft land, for which payment was made in meal, with the balance going towards eliminating rent arrears. He also offered crofters leases of 12-year periods, and he played no part in evictions.

Donald Cameron of Lochiel was another Highland laird, and was also held in some high regard by his tenants. He succeeded to his father's estates in 1858, and is on record as refusing to sanction repeated requests from his factors for the eviction of tenants or the sub-division of tenants' holdings. He played an active part in the Braes dispute as a mediator, and joined with Lord Lovat in persuading Lord MacDonald to accept a settlement which was not only in the latter's interest but in the interests of all Highland proprietors. Lochiel was in fact not unsympathetic to crofters, and his appointment to the Napier Commission was taken as a glimmer of hope that justice might be done in the end.

Charles Fraser-Mackintosh was Member of Parliament for the Inverness Burghs, and was well known for his efforts to persuade the Government to set up a Royal Commission on the crofting problem. An enthusiastic protagonist for all things Gaelic, he was a respected figure in Highland politics and regarded as one who would use his influence in favour of crofters' rights.

Sheriff Alexander Nicholson was the son of the proprietor of the Skye estate of Husabost. A Gaelic scholar, and a pioneering mountaineer (the highest peak in the Cuillin is named after him: Sgurr Alasdair), he had an extensive knowledge of the Highlands and Islands which was to become more than useful to the other Commissioners.

The final member of the Commission was Donald MacKinnon, a native of Colonsay and Professor of Celtic at Edinburgh University. Born of humble stock, he might have been a fisherman had not Lord Colonsay been struck by MacKinnon's mental agility and promise at an early age. Lord Colonsay sent him off to study at Edinburgh University, where he took a first-class degree in philosophy.

The man appointed Secretary to the Commission was Malcolm MacNeill, whose father was laird of Colonsay and brother of Lord Colonsay. He was also, however, a relative of Lady Gordon-Cathcart, of a family whose ownership of the Uists and Barra was not a happy one for the folk of those islands.

With such a Commission and its landed connections, it was no wonder that it met with an instant opposition which feared that any Report must come down heavily on the side of the landlords. But as the Commission deliberately went out of its way to hear witnesses and to study masses of documentary evidence which highlighted the defects and faults of landlordism, and, what is more, was not disturbed by the way in which the evidence was shaping, the Commission came in for favourable mentions in the radical Press.

Certainly the Commissioners did not shirk their duties, and travelled extensively round the Highlands and Islands, suffering many hardships, not the least of which was a shipwreck outside Stornoway from which all were rescued without harm. The oral evidence the Commission heard ranged from the highly articulate to the rough imperfect English of crofters more at home in Gaelic. Many crofters, in fact, elected to give their evidence in Gaelic. It can be said that the evidence now reposing in the Commission's volumes is one of the best contemporary social histories ever to be produced about the Highlands. Everything rose to the surface for intimate scrutiny: harsh dealings by landlords and factors, victimisation, starvation, abject poverty, restrictions on efforts to improve land and housing. The weight of evidence presented must at some times have seemed to the Commissioners almost impossible to synthesise into a clear picture of the remedies needed to set things right in the Highlands. There was wheat and there was chaff. But slowly the facts emerged. In general, the Commissioners were impressed by the quality of the evidence presented to them, though they noticed a tendency to flavour decriptions of the contemporary conditions with the past, with witnesses relying on '*fleeting and fallacious sources even when not tinged by ancient regrets or resentments or by the passions of the hour.*'

The Commission looked into six areas of concern: The land, fisheries and communications, education, justice, deer forests and game, and emigration. So far as the land issue was concerned, the main complaints were the size of holdings, the lack of security and tenure, the want of compensation for improvements, and high rents.

The Commission's recommendations fell under two headings: those relating to crofting townships and those relating to the status of the individual crofters. The Commission acknowledged the evidence of a strong body of opinion behind the idea that the small tenants in the Highlands and Islands had inherited an inalienable title to security on their farms, although this had never been sanctioned by legal recognition and had for long been repudiated by proprietors. While the overall claim to security of tenure could not be entertained, it was felt that security should be given to those crofters paying a rent of at least £6 a year, and who would accept improving leases. On this point, Fraser-Mackintosh advocated a lower limit of £4, to take in the bulk of crofters who would otherwise be excluded by the £6 limit. As for the thousands of cottars, who had no land at all, it was suggested they be recorded on estate books pending the time when remedies could be introduced which would '*gradually transfer and disperse this class of people.*'

The Commission proposed that the crofting townships be regarded as a distinct agricultural area or unit endowed with certain immunities and powers. The townships were to be registered to enable them to initiate improvements such as township roads, footpaths, fences and footbridges, to all of which the proprietors would make a contribution, with the unskilled labour being supplied by the crofters themselves. While the Commissioners saw that to recognise the townships was to give legal sanction to a system of common occupation with no guarantee of agricultural progress, they argued that their proposals were necessary at that stage of crofting development, as a preliminary to the time when the communal tie could be dissolved by general assent and the township lands eventually enlarged into small independent farms.

In its way, the Commission was recommending a system of rural communes, an idea which was prominent in Lord Napier's mind and probably derived from his observations of the communal life in Indian villages (he was once Governor of Madras), and the communal life of the Hebridean townships as portrayed by Alexander Carmichael, the eminent and much-respected folklorist, in a paper on 'Hebridean Agrestic Customs' published as an Appendix to the Commission's Report.

The Commission's recommendations for the individual crofter caused much heart-searching, if only to find a form of words which would give the Highland people a new lease of life, if not the start to a new future. The Commissioners concluded:

> 'We have no hesitation in affirming that to grant at this moment to the whole mass of poor tenants in the Highlands and Islands fixity of tenure in their holdings, uncontrolled management of those holdings, and free sale of their tenant right, goodwill and improvements, would be to perpetuate social evils of a dangerous character.'

The decision, then, was to give protection to a selected few who were willing and able to take on an 'improving lease'.

> 'Every occupier in a township who was not in arrears with his rent, and who paid an annual rent of £6 or more, should be entitled to apply to his landlord for an improving lease. If he failed to get it, he could make an application for an official lease to the Sheriff, who would carry out an investigation. If the application were then granted, the applicant would be given a thirty-years' lease at a rent to be fixed by arbitration, and on terms involving certain expenditure on improvements by the tenant and an obligation on the proprietor to pay for those improvements on a graduated scale. At

the end of the lease the tenant could claim renewal of the tenancy on the same terms.

Sir Kenneth MacKenzie was against giving legal status to crofting townships. What he wanted was an end to the whole system of crofting as quickly as possible, to be replaced by a system of small farms. Cameron of Locheil also set his head against the township proposals. Both these Commissioners recorded their minority dissent at the end of the Commission's Report.

The Report itself got a mixed reception. Those allied to the cause of the crofters acknowledged the Commissioners' recognition of their grievances, but felt that the proposals with regard to land tenure did not go far enough. Criticism of the proposals ranged from '*an ill-adjusted concourse of compromises and concessions*' to '*a fantastic and foolish scheme*'.

The Report also had to bide its time to come before Parliament. The Commons were at the time deeply involved with the London Government Bill, and with the Franchise and Redistribution Bill (which was to make the working class a dominant factor in British politics from then on). It was not until May 1885, after the collapse of Gladstone's Sudan policy, that the Crofters Bill was introduced to the House. Then came the fall of the Liberal ministry in June 1885, when the Irish Nationalists combined with the Conservatives in a snap vote against the Government. By December, Gladstone was back in power, but this time his own security of tenure was dependent upon the Irish Nationalists, and progress with the Irish Home Rule Bill. The Crofters Bill was again introduced in 1886 and was in due course placed on the Statute Book. But however welcome this step was, there was much uncertainty in crofting circles, for there was a considerable difference between the Napier Commission's proposals and the provisions of the new Act. The Commission's proposals for conferring a certain status on the crofting township and the granting of improving leases to selected tenants were not included in the Act. Instead of restricting security of tenure to a comparatively small proportion of existing crofts (those with rents of £6 and over), the Act extended its benefits to the great mass of crofters, irrespective of the size of holding. The main provisions were:

1. Certain conditions of tenancy were laid down in the Act, and so long as the crofter observed these he could not be removed from his croft. The more important of these were that he should pay his rent; that he should not assign his tenancy or allow his croft and buildings to deteriorate; and that he should not sublet or divide his croft without the landlord's consent.
2. Either the crofter or his landlord could apply to the new Crofters Commission to fix a fair rent, which would stand for seven years unless altered by mutual agreement.
3. Any crofter who renounced his croft or was removed from it was given the right to claim from the landlord compensation for permanent improvements which were appropriate to the croft.
4. Where land was available locally for the enlargement of crofts and this land was refused by the landlord, any five crofters could make application to the Commission for the enlargement.
5. The Act provided that a crofter could bequeath his croft to a member of his family.

Provision was made for the appointment of three Commissioners (one of whom was required to be a Gaelic speaker) to form a Crofters Commission to administer the Act. The area to be covered by the new Act comprised the counties of Argyll, Inverness, Ross and Cromarty, Sutherland, Caithness, Orkney and Shetland. '*Crofting Parishes*' were defined as those Parishes in which there were at the commencement of the Act, or had been within eight years prior thereto, holdings consisting of arable land held with a right of pasturage in common with others and in which there were still tenants of holdings from year to year who also resided on their holdings, the annual rent of which did did not exceed £30 per annum in money.

A'*crofter*' was defined in the Act as the tenant of a holding from year to year who resided on his holding, the annual rent of which did not exceed £30 in money and which was situated in a crofting parish. '*Cottars*' were provided for in the Act, but only in connection with compensation for improvements, the cottar being an occupier of a dwelling-house with or without land, and who paid no rent to a landlord.

For the first few years, the Crofters Commission was kept busy with the fixing of fair rents and dealing with rent arrears. The Commission was in fact to emerge as a quasi-judicial body dealing with all kinds of matters affecting crofting rights. In its 1891 Report, the Commission referred for the first time to the sittings by individual Commissioners as '*Courts*', and when the Commission handed over its duties to the Land Court in 1912, a substantial body of case law had been evolved and built up.

In its final Report, in 1912, the Crofters Commission made the following points, justifiably, in its own favour: Firstly, the Commissioners considered that they had left the Highland crofters in a better position than they had found them in 1886. They also stressed that the right to security of tenure and the right to compensation for improvements had produced a better standard of living, in that the 'black hovels" in which too many people had lived in 1886 had largely been replaced by smart and tidy cottages which would do credit to any part of the country. These improvements, however, were not the result of profitable land use and agricultural production. Rather, the money had come from those members of crofting families who had found work away from the home area and had remitted money back. The final word of the Commissioners was:

> 'From 1882 to 1887 the Highlands and Islands were in a state of unrest — in many places there was open lawlessness. Rents were witheld, lands were seized, and a reign of terror prevailed. To cope with the situation the Police Force was largely augmented — in some case doubled. Troopships with Marines cruised about the Hebrides in order to support the Civil Authorities in their endeavour to maintain law and

> order. The tension and excitement of those days have passed away, and peace and tranquillity prevail where 25 or 26 years ago the Queen's writ did not run.'

In 1911 the crofter came under the Small Landholders (Scotland) Act, which did not introduce any major changes, but simply extended the area within which the crofting code operated and provided wider powers of land settlement. But there was one serious defect. The 1886 Act defined a crofter as one who had to reside on his holding. In 1917, as the result of some imprecise wording in the 1911 Act, a case was submitted to the Court of Session which ruled that the residential condition was not iron-bound. This decision led to the introduction of a growing class of absentee crofters.

The Land Settlement (Scotland) Act of 1917 provided an improved procedure for the creation of new holdings and the enlargment of existing holdings under the administration of the Board of Agriculture and the Secretary of State for Scotland, which, in the course of time, saw crofting tenants occupying legally the land of large farms which had been created by the clearing of their forebears from that same land.

The next landmark in the crofting scene came when, in 1951, the Taylor Commission of Enquiry into Crofting Conditions was appointed with the remit: *'To review crofting conditions in the Highlands and Islands with special reference to the secure establishment of a small-holding population making full use of agricultural resources and deriving the maximum economic benefit therefrom.'*

The Taylor Commission's remit was concerned with the croft as an agricultural subject only, whereas the Napier Commission had also been involved with social justice, in so far as the philosophy of land-ownership in the 1880's allowed it to look at means to alleviate suffering. The Taylor Commission made a number of major recommendations, only some of which were given effect by the Crofters (Scotland) Act of 1955, which, in any case, restored to crofters in the seven Counties their own special legal code and which re-designated all small landholders and Statutory small tenants as 'crofters'.

Under the new Act, a Commission was set up to reorganise, develop and regulate crofting. It was enabled to re-let and transfer crofts; administer a scheme of land improvement grants and a scheme whereby crofters could obtain loans for the purchase of stock; it could dispossess absentee crofters whose crofts could then be re-let in the general crofting interest; it could even re-organise whole crofting townships by re-allocating the land among whose who were willing and able to work it, if the majority of crofters favoured such action. In general, however, the 1955 Act introduced little change in the legal standing of a croft.

The new Crofters Commission entered upon its task with abundant enthusiasm and made public its views on its '*special task of securing the revival and promoting the welfare of crofters.*' It also declared that nature had designed the Highland terrain '*for occupation by family units and therein lies the enduring logic of the crofting system, which relies on the participation of the family.*' The stimulation of crofting agriculture was seen as a matter of prime concern. '*........husbandry must remain the prime concern of the crofter, and his first duty is to the land so long as he holds it on a tenure of some privilege.*'

But there were problems. For instance, some two-thirds of all crofters (estimated from agricultural returns) had allowed their holding to fall into neglect because they did not depend on them for a living. Indeed, so far as many crofting families were concerned, their income was dependent upon the surplus members who had obtained work elsewhere than in the home area, so that the croft had become less and less a matter for family interest. On the credit side, the schemes for the improvement of pastures by reseeding proved a success, as a luxurious green cover took over from the predominant brown of heather. Even so, by 1961 it was obvious that an agricultural basis for the survival of crofting was no more than a heart-felt wish, and that there was a great need for ancillary activities, such as tourism, which might offer a better future. That realisation led to thinking on a new tack which was announced in 1963:

> 'Crofting has always meant the sum total of the crofter's various activities, with agriculture the common factor. In some areas agriculture may be the least productive element and there has been a steady trend in that direction for the past fifty years. The tendency to regard crofting as synonymous with agriculture must be resisted. Many crofting units must inevitably remain small either because of the nature of the land or because of social reasons. For crofts in this category there is no virtue in amalgamation for its own sake. It is a misnomer to describe the non-agricultural uses of the croft as "subsidiary or auxiliary", particularly in areas of low agricultural potential and small crofts (which is a description applicable to the great majority of crofting areas). A croft may well become viable in other ways although agriculturally it could never qualify for that description.'

This meant, in sum, that non-agricultural uses of the croft were to be recognised as part and parcel of crofting activity, which could take precedence over agriculture in importance and value, and that encouragement and incentive to create a shift of emphasis should be offered to crofters. Slowly the ideas hardened and led eventually to the submission of recommendations to the Secretary of State for Scotland for the modernisation of crofting. Among the most important of these was a recommendation that crofters should be given the right to acquire ownership of their crofts, which was seen as a means of overcoming obstacles to development and of acquiring capital for development, as well as allowing land to go to active crofters who could make full use of their holdings.

When the Crofting Reform (Scotland) Act of 1976 appeared, it differed considerably from the recommendations for the reform of tenure submitted by the Commission in 1968, even though the principle of conversion to ownership was accepted by the Government. Whereas the Commission wanted wholesale conversion of all crofts at a given date, the Government preferred to give the crofter the option of either retaining his crofting tenure unimpaired, or acquiring ownership of his croft house and of the inbye land of his croft in whole or in part.

The Act confers on the crofter the absolute right to acquire the ownership of his croft house and garden ground on payment of a price which is only a few pounds for the land where the house is wholly a tenant's improvement. If, however, the house has been provided by the landlord, or if the latter has contributed to its cost, this fact is reflected in the price payable. The crofter also has the right to purchase the inbye land of his croft, or any portion of it, at a price calculated at fifteen times the rent of the land. This is subject to a fairly limited right of objection by the landlord on the grounds of hardship or damage to estate management.

There has been no rush by crofters to take advantage of their rights. The bloodletting, hardship and suffering which established security of tenure a century ago, and the slow, almost imperceptible progress to 1976 has created an understandable caution in the mind of the crofter, and an attitude that time will show the way to go in his own best interests. The crofter, however, is provided with a number of rights which he can exercise should he feel threatened. For instance, with many Highland estates now falling into the monied hands of land speculators both foreign and British, a crofter can take action to compel the new owner to sell him his croft if he considers that the new regime might harm his interests.

It has been a long haul for the crofting population of the Highlands and Islands to achieve their present position of relative strength. The sad fact of the matter is that the land itself has not been allowed to become a significant element in the Highland economy, to provide the kind of environment which engenders a healthy social structure keen to exercise initiative and enterprise. That has yet to come.

Village Of Plocropol, Harris

CROFTING TOWNSHIP

There is a whole world of difference between the picturesque villages of the English countryside and the townships of the Highland scene, which are equally picturesque, but not so photogenic. The English settlements have a long stable history of close communal association, and also a history of regulation imposed by the local feudal manor and its lord. The surrounding countryside is witness to centuries of careful cultivation by the toiling villein and serf, and the total encompassing atmosphere is one of slow-moving, almost Utopian, contentment — or, at least, that is the impression the tourist receives. Many Highland townships, on the other hand, while some are indeed of long standing, have a much shorter history, having been created by displacement of populations in the last one hundred years or so.

Many of the present day crofting townships are products of the crofters' immediate forebears, men and women who often had to make the very soil from an alchemic mixture of seaweed and crushed stones, having been forcibly settled on the most infertile, rocky land possible. The crofting townships also have an atmosphere of perpetual change as dwellings, for instance, are re-vamped from the old style of design to the ultra-modern. There are no particular features in the township which show a development from feudal bond to freehold: no village square, no village pub to act as a neutral area for local discussion, no visible memorials to the war dead and none of the outward trappings which indicate a closely-knit social community. Yet, the crofting township *is* a tight social unit, simply because it is the folk of the township who are the dynamic elements; the physical environment is merely a stage-setting.

The landscape around most crofting townships is sparse and ascetic: therein lies some of their attraction to the tourist — the sudden, dramatic change from the familiar to the strange, almost foreign, pattern of randomly-chosen house sites. The crofting township, however, does tend to be dynamic in that changes occur almost daily, as they have done over the past few years, with the provision of such amenities as surfaced roads, street lighting, drainage, local authority housing and community halls.

It is not often realised that there are few large conurbations in the Highlands and Islands. Inverness apart, most Highland burghs have populations between 5000 and 10,000: Oban, Wick, Campbeltown, Stornoway, Thurso, Kirkwall, Lerwick and Dingwall are such. A few have populations between 1000 and 5000 people. The preference has always been for the smaller interdependent community which was socially and commercially functional, in that it was originally established to serve the people who founded it. Tradition and experience has always dictated that the smallest social unit which can constitute a township in the crofting areas of the Highlands is a team of four men with their families. Once such a team is established, multiples of similar family units created the larger settlements. Such a community is based on the provision of fundamental socio-economic needs. Its members enjoyed each other's companionship and co-operation in the types of activities pursued by the inhabitants, always tied in with the land and fishing. In the early days, the aim of these units was to create a subsistence economy, and, if things went well, a surplus for barter or trade. All domestic requirements were satisfied by activities pursued within the community.

Once the community grew sufficiently large to produce a food surplus, the specialists appeared — shoemakers, carpenters and smiths whom the community could afford to release from their basic roles as land-workers, the very roles for which they were accepted as members of the earlier, smaller, community. In this way did the specialist net-maker, weaver, mason and boat-builder come into existence. But some of those specialists required a fairly large society to absorb a year's production. If it takes a shoemaker one day to make a pair of shoes, and if one pair of shoes is worn out in a year, a community of 100 members would be too small to support the full-time shoemaker: he would be idle for most of the year. To be fully occupied throughout the year, he would require to service a community of 365 souls. Thus, the optimum size of a community was often dictated by its needs, where these needs could be provided within the native township or else within an area containing a number of townships within easy reach of each other. Later, as roads were driven out to reach those townships, there came a dependence on the externally-made product, which inevitably meant that the specialists either moved out of their community roles to revert to their original function as land-workers, or else left the district to trade in fresh and greener pastures. The theme of communal help and self-sufficiency obtained for many decades until recently in the more remote areas of the Highland region.

Very little is known about the earliest types of settlements, except in general terms. Most of the houses were either built of insubstantial materials (usually turf) or were constructed of stone and lime, where local stone would be procured easily and did not require a great deal of work in dressing. Otherwise, they were built with an inner and outer wall of rough stones, about six feet high, and the cavity between them packed with earth and clay, and the whole then topped with a thatching of whatever local material was available.

In the days when the clan chief and his clansmen were closely tied by blood and fosterage, the settlements were clustered round some kind of fortification, which gave the protection necessary for the community to survive. But for many centuries the Highlands was hardly a peaceful country, and the resultant unrest made for slow development of the settlements into formal and stable social units. Life in Highland settlements in early times must have been continually disrupted by demands for clan service of one kind or another in the neighbourhood of the clan lands and, still more, for service at a distance from home. Of course, there were periods of relative stability, but these were rarely of sufficient duration to

Old Croft Houses, West Coast Of Lewis

influence the social development of the clanfolk. The gradual alteration that took place in the relations between the chiefs and their clansmen was a modifying factor of considerable weight. As the power of the chiefs increased, the independence of the townships diminished, a shifting of tribal influence which necessarily affected the economic balance, and which resulted in social change. The pivot of the clan system was always the chief, whose function it was to act as leader in war, as protector of the clansmen, civic head, landholder and legal administrator of the clan. The most important obligation of the clansman originally was to follow his chief into battle.

For the performance of functions for the general benefit of the clan, the usual reward was a grant of land, which, over the years, tended to become hereditary. The most prominent members of the clan, apart from the chief, were the tacksmen, usually relatives of the chief, who, in return for military service, held large areas of land on lease or at a nominal rent. Most of the clansmen were servants on the holdings of these tacksmen: they were tenants-at-will of the chief or tacksmen, or else sub-tenants without any legal status. For the great majority, however, security of tenure depended on the goodwill of the chief or tacksman.

Two years after the failure of the 1745 Jacobite Rising, the heritable jurisdictions held for so long by the chiefs were abolished. This struck right to the heart of the clan system, and weakened the old military bonds between chief and tacksmen. The old relationship was replaced by one of landlord and tenant, which left the ordinary clansman at a serious disadvantage: the very existence of his township was threatened, a state of affairs which saw the eventual movement of the people out of the Highlands. Deserted groups of dwellings are today a familiar sight, silent witnesses of the history of yesteryear.

The older Highland settlements tended to be of the grouped or loosely-scattered pattern; the present-day linear distribution came with roads, or else was developed as vacant crofts were redistributed and the associated vacant houses fell into ruin. The closely-grouped township is a relic of the old Celtic type of settlement, with its significant central area for social gatherings, often at crossroads. Linear townships tend to be dead at their extremities.

It is not often realised by the modern tourist in the Highlands and Islands that many of the present-day townships, although they look as though they have been there for centuries, are in fact less than two hundred years old, and are not vestiges from times long past. Particularly where the township is situated on hard ground, surrounded by peat bog, or is restrained by a shore line, there is an indication of its beginning as a forced settlement where crofters, shifted from fertile land, had to settle where they could and create a new social unity.

The reports of the Crofters Commission in the late 19th century are full of statements about the uprooting of townships. Of Avernish, near Kyle of Lochalsh, to choose but one such example, it was said:

> 'This was at one time an important township and used to contain a large, prosperous happy and contented population, but thirty-four years ago the bulk of the people were expatriated, as usual through the agency of a factor who wished to form a sheep farm for his son. Those who were allowed by the proprietor to remain after the factor had expressed his determination to have them evicted in these words: "*Go you must, even though you should go to the bottom of the sea*", were allowed a mere fringe of the township bordering on the rocky seashore. These patches have been considerably improved since by the crofters at their own expense, and to make bad worse, some have to pay 14 shillings a year in addition to their rent for the use of seaware for their soil. When the people were deprived of their township in 1849, the land was left as a sheep farm and some years afterwards about one-fourth of it was turned into plantation.'

The factual literature on the Clearances is full of stories which tell of the way in which the old settlements were evacuated forcibly, with their inhabitants pressured to make the choice of emigration overseas or re-settlement on the worst ground available. For instance, this happened to the original communities on the islands in the Sound of Harris and on the Eastern seaboard of Harris. The folk were forced to settle on the inhospitable Western seaboard, often on rocks and clifftops, and had to make a fresh start. Pioneers in the widest sense of the word, they had to build from scratch and virtually create the soil from which their future food might grow —by ditching thin ground, breaking stones and leavening with seaweed and peat to make the familiar *'feannagan'* or lazy beds (and nothing could be a greater misnomer than to call them 'lazy beds!). Even then, they had to overcome the impositions of their landlords:

> 'Our houses are of the most miserable description and they are built so close to the sea that in tempestuous weather there is a danger that the sea may enter them. We are not allowed to cut seaware on the rocks on the shore, but have to go out and fish for the coarse kind found lying at the bottom. Often we have to fish for it with an iron hook six or seven pounds in weight, and at a depth of five fathoms.'

It is many decades now since crofting township communalism was a vital cohesive factor in the socio-economic life of Highland and Island communities. The crofting system was always one which relied on the essential co-operation by the people for the people who lived perilously near the edge of satisfaction and need. It was this fact of life which gave rise to the requirement for a minimum number of working family heads who could perform certain tasks on land or fishing out at sea. A working unit of four men was deemed the minimum to ease the progress of, say, the work of digging at a rate which would be quicker than four men working on their own and in isolation. They could also combine to form a Gaelic 'farm-team' for planting, harvesting and dipping sheep. Similarly, a 'boat-team' of four men was needed to haul a boat up above the water-line or to pull on the oars when at sea. But these teams were not purely technical or economic:

there was a vital social bond to which each member contributed for the greater benefit of the whole township, made up of multiples of four.

This aspect of teamwork was highlighted in modern times in Russia during the 1930's with the formation of the workers' teams in factories, farms and mines based on the idea of a young miner called Stakhanov. It also has interesting similarities with the new principles of teamwork in some modern Scandinavian car factories, where small teams of workers are directly responsible for the production of each individual car, a system which, it is held, increases production, gives a direct sense of responsibility to each worker, and eliminates the boredom and frustration of the assembly line. Indeed, the egalitarian aspect of the old Highland crofting regime, and that which obtained in the Soviets, cannot escape comparison, and even today the principle of social control of resources, as opposed to private control by another, is still subscribed to in the crofting counties.

The decline in crofting has gone along hand in hand with the decline of social intercourse, and while there are still often occasions where 'many hands make light work', the older element of communal interest is now rather diluted in those areas where facets of modernism have appeared. For instance, a crofter owning his own tractor has a degree of independence which is foreign to the whole concept of crofting as a way of life.

Another factor which has contributed to the dilution of the social base of the crofting township has been the necessary emigration of members of a three-generation family unit to seek work away from the native area. Even up to a few decades ago, the efficient working of the family crofting unit was dependent on all members of the family, from grandparents to grandchildren. The oldest members of the family tended to have extremely long working lives and were active long after elderly people in industrial societies had retired for a life of ease, if not enforced idleness. The break-up of the basic family unit has created a trend towards disintegration which is on-going, now that enhanced standards of living are demanded, particularly by the young, standards which are far above those which were accepted fifty years ago, say.

One cannot, of course, deny those demands for a share of the riches in the national cake, for, to be sure, a century and a half of persecution, discrimination, exploitation, famine and poverty are undoubted qualifications for some kind of compensation. Even if the result of the whole crofting exercise has been to prove that communities can survive in the face of extreme adversity, the crofting township is not yet the anachronism some agro-economists would make it out to be, for many communities still display a strength of purpose, a resolution and a powerful sense of identity which is no bad thing in these days when the remote political administrator is set on reducing people to numerators in an overall planning exercise.

Happily, in many townships of the present day there still exists a significant social bond which has always been characteristic of people whose history almost dictated that they work and sing together — that or die.

Sheep Shearing At Ness, Lewis

CROFTING TODAY

Crofting, crofts and crofters are today not what they once were. Before 1886, the activity, the tenants-at-will, and the small patches of land worked by those tenants were hardly recognised in a legal sense. After 1886, however, they were all placed firmly on the map of Scotland, and, in particular, in the seven crofting counties which then existed: Shetland, Orkney, Caithness, Sutherland, Inverness-shire, Ross and Cromarty and Argyll. A century ago there was the dire need to stabilise a worsening socio-economic situation in the Highlands and Islands, a situation which had reached the stage where direct action by crofters and their supporters had led to inevitable bloodshed. The Crofters Act of 1886 took much of the sting out of that situation. But it left the district like an insect caught in amber, and for many of the decades that followed, until 1955 when the second Crofters Act was passed by Parliament, land in many places was allowed to revert to its natural state with crofters similarly held in suspension, restricted in their efforts to make the fullest and best use of the land of the region. Only after 1955 did efforts begin to win back the fertility of the land in agricultural terms, and to make the best economic use of land that had been rendered unfit for crop and animal production.

It was against a background of eviction and land agitation that crofting was brought into legal existence. Today, with the passing of the 1976 Crofting Act, the croft has now progressed in status to something akin to a marketable commodity, in that the Act offers owner-occupancy to crofters who have traditionally claimed a symbolic right to the land over and above their actual legal rights of tenure.

Subjected as it is to many pressures, from both within and outwith the crofting areas, the special form of land use which crofting represents is now undergoing change, and from the statistics produced annually by the Crofters Commission, one can see certain trends. Each year, for instance, the number of registered crofts decreases by some 2 per cent. The average age of new tenants of acquired crofts is 41 years, while the average of those who succeed to crofts is 53. This latter trend is serious, for it indicates that the generation of young folk born and bred in crofting communities is being forced to look elsewhere for work and to take up careers which have little or nothing to do with crofting. In some areas the average age of the crofter is much higher than that already quoted. In Harris, for instance, it has been estimated at over 60 years of age. This figure, however, should be set against the fact that a crofter aged 60 is probably more fit than a much younger citizen of any urban area. Another point could perhaps be made: the older and more set in his ways the crofter is, the less chance there is of change coming from within that aged community.

Another trend which lies hidden in statistics is that while the number of registered crofts in the seven crofting counties is around 18,000, the actual number of working units is only 13,000. A 'working unit' is defined as being two or more once individual crofts being worked by one man and his family. This trend has been brought about by an increase in the number of absentee crofters, a comparatively new breed which almost negates the painful agitation which produced the 1886 Crofters Act.

Crofting land in many areas is now being used less and less for crops, as revealed by the fact that in 1973 about £260,000 was paid out in cropping grants, while five years later the figure had fallen to £37,000. However, some trends are not uniform for the whole of the crofting area. In Caithness, for instance, a higher percentage of acreage is being tilled than in Lewis. Land improvement grants payments are also highest in Caithness, and there are cash incentives for such activities as bracken cutting, the planting of shelter belts and the making of access tracks to the improved areas.

There is a wide range of grants available to crofters. Financial assistance is on offer for the growing of crops on marginal land, land improvement, the provision and improvement of fencing, drainage and other schemes such as the installation of electric generators, water supplies and the provision and improvement of roads. But it is not always possible for crofters to avail themselves of this aid. Because of lack of capital, crofters cannot find the total finance needed for desirable schemes, which have to be completed at the crofter's expense before the grants are paid. It has been said that the Treasury has a genius for making some schemes of aid quite inoperable.

The position is different in those crofting areas on which oil-related activities have made a significant impact. In Shetland, for instance, which is more of a crofting area than is Orkney, bridging loans are available from the Shetland Fund at low rates of interest to enable crofters to take advantage of the Crofters Commission land improvement schemes. Even so, and even in less fortunate areas, much land which was derelict has been taken into some kind of agricultural production. For a period some years ago, the crofters in Lewis and Shetland, in particular, and to a lesser degree elsewhere, were reclaiming some of the worst moorland in Britain at a rate in excess of 4000 acres a year.

Despite the picture conjured up in the mind when 'crofting' is mentioned, the mosaic of crofting activity varies considerably in different parts of the seven counties. There is, for example, a township in Skye, not the worst example of moribund crofting, but typical of the way in which the hard fight of last century for crofting rights has been negated. From the main road running through this township, the crofts can be seen as potentially excellent holdings, with broad acres of loamy arable land, a well-drained hill with good southern aspects, peat banks readily to hand, good water, and fish in plenty a few yards away from everyone's back door. Modern housing is witness to the availability of Government grants for improvements in the domestic aspect of crofting. Yet crofting as an activity is dead in this particular township because no beast is to be seen and not an inch of the ground is tilled. All that thrives are

Evening Milking, Lewis

Harvest, Near Eochar, South Uist

rushes, ragwort, and a forest of bed-and-breakfast signs.

In contrast one can point to the sheep stock on the island of Lewis, estimated at an on-the-hoof value of some £5 millions. In contrast, too, are some areas on the mainland which are so intensively worked that they contribute significantly to the fact that 27 per cent or so of all agricultural production in the Highlands and Islands comes from crofting activities. In these times when so many acres are being taken out of agricultural production for new housing, roads, industrial estates and the like, it is of vital importance that crofting be given its fair chance to make a significant contribution to the filling of the nation's larder. A school of thought (only one of many concerned with the viability of crofting) has advocated the re-introduction of those Government powers (used during the 1939-45 war) which insisted that land either be used properly or else placed in the hands of someone who would ensure that it was put to good use.

Some trends noticed in recent years militate against any real development towards viable crofting activities in certain areas. One such is the increase in the number of absentee crofters who make arrangements for their crofts to be worked by others, and who then often find themselves pushed really hard to make full use of their increased land holding.

> 'The whole island is gradually being taken over by a handful of crofters who seem able to snap up every croft or piece of grazing as soon as it becomes available and before anyone else has had the time or the chance to do anything about it. Most of them have as much as anyone can handle properly, yet they go on acquiring bits here and bits there. They are not doing anything wrong according to the law, far from it, but they are doing a great deal of harm to crofting, for they don't come and live in a township and contribute to its business and rear a family there. Make no mistake — we are seeing the return of the tacksmen, and that is bad news for the ordinary crofter.'

That was written in a small Hebridean newspaper as an editorial

Potato Planting, South Uist

and it highlighted what was regarded as an undesirable trend, in the context of an existing demand from young people, native to the island concerned, who would dearly like to be given the chance to make a start as a new generation of crofters in their own local area. A side effect of this particular trend is the dilution of the communal threads which are essential to the binding of a township to its common interests. The incidental presence of absentee crofters in a crofting township, who might attend meetings only to protect their own interests, could well be a disruptive one, unless, as in a very few cases, the absentee crofter, well connected in his own township, has that degree of skill and confidence which can make a significant contribution to the adopted township.

Another disturbing trend is the lack of interest in their croft holdings by many crofters who are classed as unemployed: even the production of vegetables is neglected. The reason might be that if the value of croft produce exceeds 75p per day, there can be deductions from the supplementary benefit. Many crofts are worked only to the minimum level to attract the various subsidies available. The late age of succession to crofts prevents younger and more energetic people being involved in crofting; crofting areas in general have some of the worst age structures in Scotland, being top-heavy with aged people. Crofting neither keeps young people at home nor encourages them to return while still young, something which can only contribute to a continuous decline in crofting activity.

The advocates of the crofting system say that its social aspects have always been more important than the economic benefits. But recent studies made in the seven crofting counties indicate that the social factors (which were of paramount importance in the past) are now of little importance. Ideally, the 1976 legislation, which allows tenants to buy their crofts, should result in more crofts being sold to those of the native-born younger generation, and to incomers with a definite commitment to making a contribution to the locality of their choice. This has not happened. The legislation also enables the creation of more housing in areas where there are full-time employment opportunities, so that a crofter opting to take an owner's title can then assign the croft to the new tenant, who in turn builds a house on the land. To date only in Lewis is this happening to any great extent, possibly brought about by the impact of a lucrative on-shore oil-related facility near Stornoway, which injects large amounts of direct cash into the island economy.

In the last two decades the crofter, working land which can offer only subsistence living, has been presented with increased opportunities of gaining a useful income from ancillary employment. These jobs are to be found in astonishing variety: fish-farming, forestry, construction, low-skill work on Service installations, construction yards for oil-related activities, and tourism. Many of these oportunities have appeared since the Highlands and Islands Development board was set up in 1965. The Board has ploughed much public money into creating new bases for employment, with mixed success in some areas, but by and large of benefit to the more remote parts of the seven counties. A recent innovation has been the setting up of community co-operatives within crofting areas: small, locally-based operations which pool the financial, natural and entreprenurial resources of a number of townships to improve their economic base and to create part-time and full-time employment. Most of these co-operatives are at present located in the Western Isles, but communities on the mainland are moving in the same direction. Time will tell whether there is sufficient residual enterprise and initiative in the crofting communities to sustain these developments, considering the high rate of migration from the seven crofting counties for well over a century now.

Crofting is much more than a way of life, sustained artificially and by legislation. It is still the proving ground for cultural and linguistic values, from the Gaelic-based domain in the west of the Highlands to the characteristic Norse-oriented domain in the Shetland and Orkney Isles. It is a storehouse for residual moral and social values from which the nation as a whole can draw when these become debased in anonymous urban societies. It is also a system which holds a population on relatively poor land which would otherwise revert to an irretrievable natural state were it not being tended as it is at present.

In 1967 the first Report of the H.I.D.B. said of crofting:-

> '....if one had to look now for a way of life which would keep that number of people in relatively intractable territory, it would be difficult to contrive a better system. But its future depends on other employment and support. This the Board accepts as a clear challenge and duty.'

This view has been supported by the Crofters Commission:-

> '.....the crofting system has considerable social, cultural, linguistic and economic importance at the national and indeed European level. It would be a very great loss —and would have almost unthinkable consequences— if the system were to be threatened seriously by artifical inducements to amalgamation in the pursuance of structural improvement.'

But it may well be that the crofter does not see himself as the curator of a linguistic culture and keeper of a form of land use which offers only a subsistence level of living standards unless it is supplemented by suitable part-time occupations. He no doubt realises that the vast majority of crofts are too small to provide an adequate family livelihood. And he also recognises that the available part-time employment is often incompatible with the demands of crofting. Yet, that is the system which is being perpetuated. While, too, the crofter may feel protected under existing crofting legislation, such legislation is in no way immune from Government interference at some time in the future, for it could be modified or repealed in the same way as any other Act on the Statute Book. Then there is the problem of those who possess the land, who often acquire land as a hedge against future bad times and who are, too often, perfect strangers to the whole history of crofting. They may well turn out to be the bringers of change so drastic and dramatic that the skin of the Highland crofting bubble may be pierced with as much pain and distress as was seen last century.

Land, in fact, is the key to the future of the Highlands and Islands, in socio-economic terms and not only in the context of crofting. History has shown that the region can support forms of land use which approach an ecologically acceptable compromise, where natural resources and the human element exist on a common plane for the assured survival of both into the future. History has also shown that certain form of influence produced by such abstractions as politics and the materialised aspirations of non-Highland interests can quickly destroy the economic viability of the land as created by those whose very lives depended on it.

Land, in fact, is the perennial issue, of which crofting only forms one socially significant part. The existence of the Crofters Commission has merely served to surround crofting as a form of land use with an aura of protection, a kind of force-field, against which all attempts to introduce real change from outwith the region are parried, and attempts to create new conditions for relevant change from within are thwarted. The Commission itself has said as much:-

> '.....there is the risk that the existence of the Commission as a completely separate body will create the illusion that the crofters' interests are specially protected, whereas in fact he may be less well served than other classes in the community because of the peculiarities of crofting tenure and the division of responsibility for his welfare among so many different bodies.'

In view of the seemingly endless powers vested in private land-ownership, it might seem strange to mention that the land of the Highlands and Islands can actually be taken over under the powers vested in many existing clauses in different kinds of legislation passed by Parliament during the last three decades. The Secretary of State, the Highlands and Islands Development Board, and a mooted Rural Development Board, can all acquire land by agreement or compulsorily, depending on the circumstances obtaining at a particular time. *(See Appendix.)*

The future of crofting and the crofters thus lies firmly embedded in legislation. Only human will, urged on by the best of motives, will ever prise it out of its statutory stagnation for the benefit of those whose attitude to the land of the Highlands has been proved to be the best one in the past. That past could live again.

Thatched House, North Uist

HOUSING

The Gael has always been an outdoor man. 'Home' to him was the great outdoors, and his house was merely a convenient shelter from inclement weather. It was not an object of domestic luxury. Rather, it was a building erected to shut out the storm. In good weather it was normal to live out of doors, for instance during the summer months when the folk of the older Highland townships went off to the hill grazings, to the sheilings, which were something of an extension of the home. James Boswell, Dr. Samuel Johnson's companion, describing his visit to Coirechatachan, remarked: *'We had no rooms that we could command, for the good people here had no notion that a man could have any occasion but for a mere sleeping place.'*

The house or dwelling place was, however, a focal point of some social significance, as indicated by the Gaelic expression *'aig an teine'* (at the fire), applied to the main room of the building, and suggesting the value attached to the social graces and aspirations of friendship. When the exiled Highlander speaks of 'home' he rarely means his parents' house; rather he means his native village, and, in the larger context, the immediate area of moors, hills, glens and rivers which made up the place of birth. This attitude meant no disrespect to the dwelling where one's family had been born and brought up. But it was ever the case that it was the natural and physical environment, the wider elements of sky, moorland, hills and the like, which provided the basic identification. This is something which may well stem from the old times when a clan had its identity tied in closely with a well-defined territory.

In these modern times, when planning consent must be obtained for nearly everything, if a crofter builds a new house, the older house must be pulled down. But before the advent of the planner, when a new house was erected the old place was merely put to new use as a barn or store, and was seldom razed. In some cases the old home was left to the rigours of wind and rain and allowed to fall into a tumble of stones, to become some kind of visual reminder of a former time and style of living. In many places in the crofting areas of the Highlands and Islands one sees the older dwellings, now softened by time and weather, standing beside the new structures: neglected and forlorn — but not forgotten.

It was this attitude of the Highlander to his house which was often misunderstood and misinterpreted by visitors to the Highlands a couple of centuries ago, who stood aghast when they saw the conditions in which they found the local people living. Yet such conditions were little different from those found in other parts of rural Britain and, indeed, on the continent of Europe. The Highlanders' homes were often described as dirty and dark. Some were likened to '*smoking dunghills*' with interiors exuding '*nastiness and simplicity*'. Thomas Pennant, on Skye in 1772, visited the house of a piper by the name of MacArthur, the hereditary piper to the Macdonalds, and wrote:-

> 'His dwelling, like many others in this country, consists of several apartments; the first for his cattle during winter, the second is his hall, the third for the reception of strangers, and the fourth for the lodging of his family; all the rooms within one another.'

The last observation was meant to convey that these rooms were all contained in one structure and were under the same roof.

As might be expected, considering the large area covered by the crofting counties, the basic dwelling, popularly called the '*black house*', varied in its design to conform to local climatic and environmental conditions. Thus, the Lewis or Hebridean type of house differed in some essentials from that found in Skye and on the mainland of Argyll, particularly in the construction of walls and the type of thatch.

The Hebridean building was generally an oblong structure, which varied in length according to the means and requirements of the occupier. It had but one door and frequently no windows. Early last century chimneys were not favoured, as any opening provided for the escape of peat smoke tended to reduce the quantity of soot deposited, and soot was regarded as a valuable fertiliser. A concession for the letting in of daylight was often a single pane of glass fixed into a hole in the roof. But that, too, soon had a coating of soot and the original purpose was defeated. All the sun and daylight was admitted through the open doorway. Both family and cattle entered by the same doorway and lived under the same roof, an arrangement which allowed a significant and convenient accumulation of cattle dung. A description, c. 1830, of an old Lewis thatched house runs:

> 'If one entered such a house in the month of May, after the crops had been sown and the manure cleared out, the visitor would have to descend a foot or more from the level of the doorstep, thence onward towards the portion of the house occupied by the family, when he would have to step up a foot or so to reach the level of that floor. Later on in the season, the visitor would find that the cow dung, to which a considerable quantity of sea-ware and earth had been added, was on a level with the doorstep. Towards the beginning of spring, the manure heap rose considerably above that level, and the visitor would have to get to the top of a plateau, and thereafter descend into the family circle.

When the spring tillage began, the manure was carried away in wicker baskets or creels to the arable land, or, if the tenant had a horse and cart, the gable end of the house was pulled down and the cart backed in, loaded up and driven away. The operations liberated noxious gases from the decomposing mass which only those accustomed to them from their youth could bear. Even so, there were often many victims falling ill to the effects of 'dung-fever' and ammonia fumes, which claimed their heavy toll. It was this problem which the owner of Lewis tried to eliminate in 1830

Roof Repairs With Grass Turves, Lewis

when he issued instructions that a partition be erected between humans and cattle, *'and that more light should be admitted into the dark recesses of their habitations.'* In several instances the new rules were complied with, *'but sorely against the wishes of the people'.*

The common method of construction used for the walls was a double skin of large, undressed but reasonably regular stones, set some few feet apart, and the inner gap filled in with soil and rubble, compacted to form, in the whole, a structure of superior strength, able to withstand wind and rain, and which afforded considerable warmth to the family and animals inside. In the older Hebridean type, a broad ledge often ran round the perimeter of the outer wall, on which grass was allowed to grow to provide a tempting morsel for sheep. In the Islands, the thatch came down as far as the central core of rubble, while on mainland houses the thatch protruded over the outer walls.

The roof timbers were often large spars of driftwood, obtained from the sea shore, from the wrecks of wooden sailing ships which met disaster in the turbulent seas on both sides of the Hebrides and on the west mainland coasts. In many places in the Highlands the timbers were the property of the tenant and regarded as a moveable asset. It was this fact that led to the burning of roofs during the Clearances, to prevent an evicted tenant attempting to set up another structure elsewhere. A Report of 1850 reads: *'His capital consists of his cattle, his sheep.....the rafters of his house.....'*

The thatch was composed of whatever material was available, and laid over a layer of sods, put down from eaves to ridge and overlapping each other. Straw, rushes and bracken were all used, variations being struck according to what was available in the district. The thatch was kept in place by ropes weighed down by large stones. Straw ropes and heather ropes were often used when the hemp type was unavailable. The floors of the buildings were mainly earthen, though flagstones were popular for the living areas. One traditional floor consisted of earth, mixed with white sand, the whole trampled down hard by a flock of sheep kept on the

Carnish, North Uist. c.1880

move for a whole day and then let out. When this was swept clean, the result was a hard-packed surface that only required a dusting of sharp sand to redd up and renew.

The internal walls were usually left as they were built, in natural stone. But it was quite common to mud-plaster the walls to obtain a smooth surface which could then be covered with wallpaper, paint or limewash.

The fire was often set on a raised slab in the middle of the living room, with the pots and kettle being hung from a long chain, called a '*slabhraidh*' fastened to the roof-ridge. Only later was the fireplace set in a gable end and a chimney provided. Apart from cooking food, the fire played an important role in keeping the whole building warm, to prevent the sods on the roof becoming waterlogged and so heavy that the rafters might collapse. Contrary to popular belief, the interior of the house was not so smoky as to be uncomfortable. Peat reek tends to rise and the air is relatively clear near the ground. Chairs and stools were usually low-set to enable the folk to sit below the canopy of smoke trapped in the recesses of the ceiling. It was the unfortunate visitor who was offered a higher seat, and who found the need for fresh air quite compelling — as Prince Charlie discovered when he landed on Eriskay and was offered shelter, accommodation and the hospitality of an old black house.

In the older type of croft dwelling, the furniture provided was basic. Alexander Smith, author of the popular 'A Summer in Skye', says:

> 'The sleeping accommodation is limited and the beds are comprised of heather and ferns.....the furniture is scanty; there is hardly ever a chair — stools and stones worn smooth by the usage of several generations have to do instead.'

However, that was at the lower end of the social order. Other houses possessed standard items of furniture such as box-beds, a bench or settle, a table, a dresser, chairs, stools and chests. Many items were home-made and often advertised the fact. However, as pretension and the need to show off was generally far from the mind of the ordinary folk, so long as the item was functional, that was all that really mattered. Then, as it is today, some of the most comfortable armchairs are those of many years' standing.

Other articles to be found included the inevitable tools of the spinning trade, cards and spinning wheels, and those used for the storing, preparation and cooking of food.

Artificial light was provided by an iron '*cruisgean*', one or two shallow pans filled with fish oil into which the tail of a wick of dried pith of rushes was dipped. Candles were a later provision, before the advent of the oil-lamp with the tall clear glass chimney, giving off a cosy wonderful yellow light. Much later was the 'Tilley' pressure lamp, which provided light by means of an incandescent mantle, to say nothing of a stream of warmth which could fill a room quickly.

Actual domestic items included tubs for water, milk and sour milk, with often a fourth tub in which milk was placed to curdle for the making of crowdie or soft cheese.

Once the croft was made a legal entity and the crofter given a Statutory status under the 1886 Crofters Act, improvements to housing were carried out according to the means available. Even so, in the Hebrides as late as 1893, recommendations by Sanitary Inspectors were still being made that the partition separating human from beast be taken up to the ceiling:

> 'In regard to the dwelling-houses it was agreed to give notice to the people by a circular to each householder, that the Local Authority will insist henceforth that a wall of stone and lime reaching to the roof be built in each dwelling house, separating the cattle end of the house from the other portion, with no internal communication and that each end be provided with a separate entrance from the outside, and that everyone failing to carry out this regulation is liable to be summoned before the sheriff.'

Whatever improvements were carried out in and around the house there were still problems with sanitation and water supply. In this respect it took many of the old County Councils literally decades to provide proper services to many rural areas, the excuse always being the low rateable values of the crofting areas hampering the generation of sufficient money to apply to such much-needed schemes. As recently as twenty years ago it was common for water to be drawn from one or two township wells, a fact which was often highlighted in Gaelic lessons of the time, when the phrase *'Tha mi a' dol do'n tobar'* (I am going to the well) was as relevant as *'My postillion has been struck by lightning'* was in its time last century.

Croft houses now tend to be modern and contain many aspects of a high standard of living, all complying with the requirements of Statutory legislation. Even older types of houses with traditional thatched roofs, such as may still be seen in South Uist, are well appointed to offer a high degree of comfort to their owners. Running water and serviced drainage schemes have now added to their amenity. In many ways, those modernised 'originals' offer the true link between the 'quaint' dwelling-house of yesteryear and the present day, though it has taken something over one hundred years to achieve.

In 1976, the Crofting Reform (Scotland) Act conferred on both crofters and cottars the right to buy their house and garden ground for a nominal sum, and to acquire the inbye land at a price calculated at 15 times the annual rent. In this way the security of the crofters' tenure has been enhanced, though embedded in a Sargasso Sea of complex regulations.

Spring Digging, Ness, Lewis

LAND USE

Land use can be defined as the retention or occupation of land for any purpose which does not exhaust its mineral or biological resources. In the Highland context, the positive aspect of that definition has had an unhappy history, created by a wide spectrum of circumstances, all man-made and often reinforced by local or national political aspirations and brought about by the philosophies which surround the possession of vast personal wealth. Highland history has created the pattern of land use which obtains in the region today. The land has deteriorated in fertility. And in some areas there is such a degree of social dereliction that it is doubtful whether recovery is possible. The Highland past has also had its effect on the basic ecology of much of the land in the region, so much so that the cost of re-instating derelict land would have to be a long-term exercise undertaken by one generation for the benefit of its successor. And even if that were undertaken, it is questionable whether the Highlands in the future could ever match the viability which existed in the past, when a reasonably large population had the potential to live in relative comfort and ease.

Up to the seventeenth century, land was held in clan or in individual peasant holdings. Vast areas were covered with the original Caledonian forest of birch, juniper, alder and scattered pine, with hardwoods in not a few places. Those woodlands gave cover to the red deer which was then a woodland animal and not the open moorland beast it has been forced to become. The deer shared its natural habitat with such natural predators as the wolf, and such domesticated animals as the old Highland black cattle and the goat. In general, the land was productive, in that in most years it met the basic requirements of the self-sufficient communities which were based on an interdependent social union where the ties between the community members were by blood or fosterage.

During the seventeenth century, large tracts of land were made the subject of royal gifts to individuals for services rendered or for expressions of loyalty made manifest in the provision of fighting men. Forest land was cleared to provide fuel for the domestic needs of an increasing population and for such industrial enterprises as iron-smelting and for the ships of the British Navy. In a more pioneering society of the type which found itself in the Americas, this cleared land would have been put to good agricultural use. Instead, the land was allowed to become infested with heather and bracken, so much so that the red deer had to adapt itself to an open environment, and, in the process, became a smaller animal than its forebears had been. In addition, due to the generally poor state of knowledge of agricultural techniques, much land was deprived of its fertility and local concentrations of population exhausted local natural resources.

By the eighteenth century, much of the Highland area was in the hands of a few powerful individuals. The original so-called clan lands, held by the chief as representative of his clansmen, became his outright possession in law, a change of status which was accompanied by a widening rift between chief and clansmen, who were now tenants-at-will, and the 'will' was that of an increasingly remote and de-cultured clan chief, whose major requirement had become money, rather than fighting men. The system whereby land was worked and used for the general good was fractured by small-holders being cleared off land in forced emigration schemes to make way for the creation of larger economic holdings for raising the sheep flocks brought up from the Scottish Borders. The prime product of the Highlands, the black cattle, disappeared as the habitat tried to support the grazing habits of two discriminate and incompatible animals. Inevitably, a coarser vegetation took over, and in attempts to improve matters, the graziers took to heavy burning. But this had an adverse effect on certain types of soil, and basic soil nutrients rapidly disappeared. The result of the burning policy was affluence for a few, a state of affairs which lasted until the soil once again failed to overcome a shortfall in grazing land. The coming of sheep to the Highlands had indeed created a desert.

In the nineteenth century, Highland landowners were extremely wealthy from fortunes made largely in industry and in speculative forays in railways and the increasing industrialisation which was to blight the country and herald urban slums. With the decline in sheep prices, the land was given over to supporting deer, a change which increased the deer forest area in the Highlands by as much as ten times in half a century, and Highland economics became based on the sport of deer-stalking. The following century saw many large estates being broken up and sold as package deals in smaller lots, to create a hierarchy of numerous petty landowners, resident and absent, who were both private and institutional. The State also became a Highland landowner after the First World War, with the State settlement of crofters and the acquisition of land by the Forestry Commission, which began a programme of plantings on land which had once supported sheep grazings. Crofting as a form of land use was in the meantime stifled by the Crofting (Scotland) Act of 1886, which gave the crofter a legal status and security of tenure, but allowed little scope for enterprise and initiative to bring the land back to its former fertile condition.

The land problem in the Highlands today has two stark faces. One is the land itself, impoverished by the spread of heather and bracken, soured by the free growth of peaty podsols with a high acid content. The other side of the problem is human: a low density of population, with little incentive to tackle the task of socio-economic regeneration and, indeed, a lack of financial resources to carry out the job even if those folk with initiative and entreprenurial skills were readily available.

The right to possess land has created political problems in countries furth of Scotland, and solutions have ranged from outright purchase and division among pioneering groups, to bloody revolution. In the Highlands, attempts at solutions have been only token. Ownership of Highland land is today as much a contentious subject as it was one hundred years ago when the Highland crofting community mounted a campaign of activity which, bloody though it turned out to be, was only partially resolved in the provisions of the 1886 Crofting Act, but which left the problem of ownership and the rights entailed in land possession on a shelf, where it still lies today, being dusted off occasionally when the subject is raised at political conferences and election hustings.

About ten percent of all Highland land is in State ownership, of which half is State forest holding and half agricultural settlement. The remaining ninety percent is in private ownership. Indeed, in real terms, one-tenth of one percent of the Highland population own two-thirds of the Highland land mass. In Caithness, less than twenty people (or companies) have legal possession of that county. Less than forty owners possess eighty four percent of Sutherland. Rather more people (about seventy) own eighty percent of Ross-shire. Another interesting fact is that many of this tiny minority of landowners are descendants of those who were in direct opposition to the Land League of last century. It is a reflection on those of crofting stock and with crofting interests that the crofter has not progressed much since 1886, that the blood and tears, the hardship then willingly borne to achieve a goal, that the terrible pages of history which had to be written to gain a place on the Statute Book of the Mother of Parliaments — that all these have not been sufficient inspiration for activists to take the issue of Highland land ownership into new and contemporary political arenas.

At the present time, less than ten percent of the land area is devoted to crops and managed grassland. Rough grazings, much of which can be classed as deer forest, account for seventy percent. The balance includes land used for forestry and moorland for grouse shooting.

The crofting element in this picture of land use, taking in arable land, grass and common grazings protected by Statutory rights, amounts to less than two percent of the Highland area. In general terms, considering the poor soils and difficulties of climate, the production from crofting activities is no less in terms of efficiency than that found in other forms of agriculture; indeed, taking into consideration the special factors with which crofting has to contend, it is even greater. Crofting of itself, however, due to the small size of the average crofting unit, has never been a viable agricultural activity. Over half the croft holdings are less than ten acres of in-bye land. On the other hand, it has been openly accepted that crofting activity on marginal land which might otherwise remain unproductive has been a major factor in retaining what little population the Highlands has had over the years.

While the ravages of history might well take a significant share of the blame for the present-day dilapidated condition of Highland soil, the methods used to try to whip the land into productive submission, by both landlord and crofter, have often been called into question, even though the latter has made amends in the reclamation, by re-seeding, of moorland. But the acreages won in this way do little to offset the creation of wastelands by burning. The destruction of the original forested land, accelerated in the 18th century by the demands for fuel, ship-building and iron-smelting industries, destroyed a natural habitat and covering which was self-perpetuating. Once the sheep regime was established by huge flocks of that 'white scourge', the decline in numbers of cattle inevitably led to deterioration of the grazings on the middle heights of the hills. The system of transhumance had been an important factor in the process of soil amelioration in relation to the underlying rock. Frank Fraser Darling has written:-

> 'The coming of the sheep caused a revolution in natural history quite as certainly as it altered the lives of the people of the Highlands, and the subsequent history of the people has been quite definitely affected by that revolution in natural history and the new eco-systems it brought about.'

After the de-forested areas had provided flockmasters with succulent grazings for some decades, the decline in land fertility began — and with a vengeance. Ground fauna began to disappear: earthworms and wood ants became extinct local species. Heather became less tender and an advancing rough herbage floor became less attractive to the sheep. The grazing techniques of the sheep, selective nibbling and pulling, soon began to show in grazings reverting to coarse land to which the flockmaster applied his only solution: muir burning. This practice can be accepted on certain types of terrain and, indeed, has been in use for many centuries in the Scottish lowlands. But the practice, as it was applied in the Highlands, was a disaster. Burning reduces dramatically the range of flora to a few specific species. It also lays soil bare and exposes it to climatic erosion. Peat, if burned continuously, develops a thick impervious skin which sheds water, and if this is in sufficient quantity,it drains away only to create bogs, marshes and water-logged ground, and leaches out of the soil any nutrients which might otherwise have survived intact in the former times when the ground maintained a woodland cover.

Dr. Frank Fraser Darling, a conservationist far ahead of his time, wrote in the early fifties:-

> 'The continuance of heavy sheep-grazing and burning for 200 years, on ground previously growing woodland and savannah and grazed predominantly and only seasonally by cattle, has changed the 'soil' of much of the West Highland hills from a friable mould with some mineral particles, to a tough, dense, rubbery peat, which serves as an insulating layer preventing access to any basic material nearer the rock. It is so acid and inactive, as well as damp and cold, that decomposition of the plants growing on it does not equal the year's growth minus what is eaten by the sheep.......Often enough, under the irresponsible conditions under which burning is all too general in the West Highlands, the peat itself catches fire in very dry springs and areas of absolute sterility are created which are thereafter particularly open to erosion by wind and water.'

With such a legacy, so painfully visible today, it is little wonder that the twin issues of land ownership and land use have occupied the fertile minds of those who, while admitting political aspirations, have been determined to use political channels to improve the situation, even to the extent of agitation for community ownership of land as opposed to State ownership. And only very cautiously have successive Governments and their continuing agencies responded by testing the water, taking such an intolerable time that the crofter may well become an endangered species before a final solution is reached.

Land use in the Highlands is an area in which the atmosphere is highly charged with political considerations, an historical legacy which militates against attempts by members of the crofting community to make better use of their land. Incentive is not encouraged, except insofar as it is concentrated on areas of land which do not impinge or encroach on the large tracts which are tied up for private use. Land is one of the main natural resources of the Highlands. Its well-being is vital to the people now living there if that land is ever again to help support a healthy social structure with an accepted economic viability.

From this general picture of land use in the Highlands, it is instructive to look back to former centuries to see how land was used, other than for the grazing of cattle and sheep. Records show that, despite the unsettled history of the region, many communities were able to put down sufficiently strong roots to develop into stable societal units and use what land they had for agricultural produce. Early agriculture was based on a system of intermixed strips of land which were periodically allocated among groups of joint cultivators, with nearby pasture grazed in common. The system was called 'run-rig', from the Gaelic '*roinn-ruith*', though other names were used in the Gaelic-speaking Highlands. The annual balloting for these strips of cultivated land ensured that no one person got bad ground continually. While the system was acceptable within the context of subsistence living standards, it did not allow for enterprise and iniative, nor, indeed, for the proper care of ground, for who would bother fertilising a patch of land which might not be his the following year? Even so, the system was compatible with the needs of former generations, and the fact that it sustained communities in basic produce indicates its then contemporary viability. Only when the population began to increase did the system fail, to fall under the attention of such 18th century agricultural improvers as Sir John Sinclair, who advocated the introduction of new agricultural methods to achieve higher returns from the land.

In those areas, such as the north-west coastal areas of the Highlands, in Shetland, and in the Western Isles, which are not blesed with deep and fertile soil, the method used was '*feannagan*' or 'lazy-beds', which in fact were far from being lazy, nor was the description a fair one of the crofters who tilled those man-made ridges of soil.

Where the soil depth was shallow, it was essential to obtain some degree of depth by piling up ridges of earth (more often than not it was peaty soil), and manure them with seaware brought up from the shore. The ditch between the beds allowed for drainage. Indeed, this is the only way in which such shallow soil can be put to use. The produce of the beds was a vital factor in the continuity of communities which, apart from diseases such as potato blight, were able to survive on a year-to-year basis. Today, all over the Highlands and Islands, there is visible evidence of the former use of land in the ridges now overgrown, but best seen when the sun is setting, when long shadows pick out those '*feannagan*' like ghosts from the past. The agricultural implements used for those ridges of soil were developed to suit the special conditions. The *cas-chrom,* the *flaughter spade*, and the *croman* were all designed for the individual working on his own, though as part of a team.

Nowadays, the old laborious processes of winning over the confidence of the land have gone, and have been replaced by the ubiquitious tractor and chemical fertilising agents. The social aspect of crofting agriculture has also changed, particularly in those areas where the crofting activity is supplementary to the cash earned by other work.

Only when the Highlanders and the people of the Islands themselves are in control of their own destinies will the Highlands and Islands begin again to make their unique and significant contribution to the socio-economic fabric of the nation. And only then will there be a regeneration of confidence in native ability to work out their own destiny for themselves.

WHATEVER HARVEST REMAINS

There have been few periods in the history of the Highlands and Islands when it could be said that life was comfortable, when times were settled, and when neighbouring clans temporarily ceased their feuding to attend to the basic necessities of providing food, a roof over the head and a fire in the hearth. Time and again the subsistence level of existence was high-lighted by bad harvests, too often followed by famine and a dearth of even the minimal requirements to keep body and soul together. In 1634, the Privy Council in Edinburgh considered an appeal from the Bishops of Caithness and Orkney for food:

> Multitudes die in the open fields and there is none to bury them.....the ground yields them no corn.....and the sea no fishes.....some devour the sea-ware, some eat dogs, some steal fowls. Of nine in a family, seven at once died.....some have desperately run in the sea and drowned themselves.'

The problem of famine or scarcity was not, of course, always a Highland one. It could be a matter for national concern. An Act of 1615, issued by the Privy Council, actually forbade the export of eggs from Scotland, for fear that this food source might be so depleted that '*in a very short time there will be no eggs nor poultry be funden within the country.*'

In 1772, the people on Skye found themselves reduced in cicumstances to such a degree that they were living on the carcasses of cattle which themselves had died of hunger. A decade later, the combination of severe storms and a crop failure brought a period of renewed hunger, a situation which was alleviated by the British Government sending food relief to the value of £17,000 into the Highlands. This food was in fact peasemeal obtained from stocks surplus to the Army's needs. That year, 1783, is now remembered in Highland folk memory as *'bliadhna na peasrach'*, the year of the peasemeal.

The plant which was to make such an impact on the Highlands was the potato. As far back as the 1680's, according to Martin Martin, the potato was a major part of the common diet, though it had taken some time before it was commonly grown in the crofting areas. It was not until the later decades of the 18th century that it really became king, because it was tolerant of a deficiency of lime in the soil, and it reacted well to the use of seaweed as a fertiliser. By 1811, it was estimated that the potato constituted 80 percent of the crofters' diet, which was seriously deficient in Vitamin C. But the very dependence on the potato was to create a problem. A report issued during the famine in Ireland and in the Highlands in the 1840's stated:

> '......those who are habitually and entirely fed on potatoes live upon the extreme verge of human subsistence, and when they are deprived of their accustomed food there is nothing cheaper to which they can resort. They have already reached the lowest point of the descending scale, and there is nothing beyond but starvation and beggary.'

The spring of 1846 was a mild one. It was followed by a warm summer which augured well for the crops, with the promise of a good harvest. But in July, the rains came, and with the rain came the blight which the year before had devastated much of the European crop. By the end of July and August, the blight had done its cruel work in the Highlands. A reporter of the *Inverness Courier* wrote of Skye, Lochalsh, Kintail and Knoydart:- *'In all that extensive area,* (he) *had scarcely seen one field which was not affected — some to great extent, and others presenting a most melancholy appearance as they were enveloped in one mass of decay.'* Everywhere there came a *'foetid and offensive smell which poisoned the air.'* In Skye, the blight struck like lightning: *'In the course of a week, frequently in the course of a single night or day, fields and patches of this vegetable looking fair and flourishing, were blasted and withered and found to be unfit for human food.'*

In Skye that autumn, less than 20 percent of the crop, planted with such hope at the beginning of 1846, was fit for harvesting. The degree of loss varied considerably. On the far northern coast the crop was almost intact. But the harvest in those districts which escaped did little to offset the disaster represented by the failure of the crop as a whole over the crofting counties. By the end of 1846 it was estimated that at least three-quarters of the entire crofting population of the north-west Highlands and the Hebrides were completely without food. Only those with access to other sources of nourishment (the sea, the cliffs, the seashore and shellfish), and those who had the practice of growing some other crops, were able to alleviate the prospect of stark starvation. But in particular those with no land were *'reduced to a state of abject famine, and live the most part on seaweed and scanty supplies of shell fish.'*

As if this were not enough to contend with, the winter of 1846 was one of frequent gales, storms and snow, and that added its contribution to the general misery felt by much of the crofting population. Continued cold and hunger produced sickness and disease. Cholera and typhus broke out. Seafood, particularly shellfish, failed to satisfy the pangs of hunger and produced dysentery instead. To cap all this, it only needed the crofters' dependence on the potato in previous years to reveal a serious lack of Vitamin C and pave the way for an outbreak of scurvy, an ailment which had not been present in many Highland districts for well over a century.

The amount of human suffering was appalling, even in the eyes of observers who might have been hardened by scenes of war.

Many descriptions of the scenes beheld by visitors and researchers working for the various relief bodies are almost beyond belief. One such is from a report made by a member of a deputation sent to Skye at the behest of the Free Church. The visit was made to the parish of Strath, in Skye, on Christmas Day, 1846:-

> 'We found the condition of very many of them miserable in the extreme, and every day, as they said, getting worse. Their houses — or rather their hovels — and persons the very pictures of destitution and hopeless suffering. A low typhus fever prevails here in several families who semed to be left to their fate by their neighbours. In one most deplorable case, the whole family of seven persons had been laid down, not quite at the same time, in this fever. The eldest of the children, a son about nineteen years of age, had died just when his mother was beginning to get on foot. No one would enter the house with the coffin for the son's remains. It was left at the outside of the door, and the enfeebled parent and a little girl, the only other member of the family on foot, were obliged to drag the body to the door and put it into the coffin there, whence it was carried by the neighbours with fear and alarm to its last resting place.
>
> When I entered the wretched house....I found the father lying on the floor on a wisp of dirty straw, his bedclothes, or rather rags of blanket, as black nearly as soot, his face and hands of the same colour, never having been washed since he was laid down; and the whole aspect of the man, with his hollow features and sunken eyes, and his situation altogether was such as I had never beheld before. In a miserable closet, beyond the kitchen where the father lay, I found the rest of the family, four daughters from about eleven years of age to seventeen, all crammed into one small bed, two at one end and two at the other. The rags of blanket covering them worse, if possible, than those on the father; their

> faces and persons equally dirty, the two youngest having no night clothes of any kind. One of these poor girls was very ill and was not likely to recover. The others had the fever more mildly, but had not yet been so long in it. The effluvia and stench in this place, and indeed in every part of the miserable dwelling, were such that I felt I could not remain long without great risk of infection, as there was no means of ventilation whatever, and not even of light. The poor woman said she had got a stone or two of meal, she said she did not know from whom, which had barely served to make gruel for the unfortunate patients. The family had no means whatever of their own.'

Of the crofting communities with access to the seashore and the sea, only those whose economic bases were partly derived from fishing were able to look to that resource for sustenance. For the most part, crofters were in no position to look to fish as the obvious substitute for the now-failed potato. The fishing industry at that time was, in any case, largely in the hands of east-coast fishermen who reaped the benefits from the silver harvest and left little for the indigenous population but the summer shoals of herring which appeared at random in sea lochs. A Government enquiry brought out this fact in one of many Reports that were to be produced on the crofting population in ensuing years. Regarded as *'ostensibly a fishing population'* the crofters were *'so poorly skilled and so ill provided with means that.....they did not possess the ability to pursue the occupation with effect.'*

The alternative to starvation was migration to the south, to find food if not work. Thus it was that a steady stream of people left their homes, often for good, to live and work in the strange new environment of, for example, railway construction.

When it became apparent that the circumstances surrounding the potato famine in Ireland could be repeated in the Highlands, Government officials were alerted by those who foresaw *'a very unusual calamity'* as the Marquis of Lorne described it. And for once the Government acted with alacrity. It could afford to, for it already had the experience of dealing with the loss of the Irish potato crop, and it had the man to deal with the problem: the Treasury's Assistant Secretary, Sir Charles Trevelyan, who responded by sending Sir Edward Pine Coffin to the north.

Coffin undertook a four-week voyage of investigation along the Highland west coast and to the islands, and saw enough for himself to confirm the deluge of reports reaching Edinburgh: that the crofting population needed a massive relief programme if thousands of people were not to die of starvation. A first reaction to the impending disaster was that Highland landowners should be made responsible for the relief of their tenants. But they responded by pleading insufficient resources. Coffin remarked that this attitude indicated that the '*moral obligation supposed to attach to the landowners cannot be relied on to secure the people from destitution.*'

It did not take Coffin long to get things organised, if only to make a start at tackling the problem. Two mills were set up to begin grinding meal 'night and day', and two frigates of the Royal Navy were converted to act as depot ships to carry meal to Tobermory and Portree. But if those who were on the point of starvation expected free handouts of food they were mistaken. It was decided that all food resources in an area had to be completely exhausted before the meal was to be distributed. Not only that, but a 'fair price' was expected to be paid for the meal. In the event, Coffin found that the old idea of free trade allowed a number of entrepreneurs to make handsome profits even when they sold meal at the prices being charged at the Government's meal depots.

Coffin's task was made more difficult by the fact that prices had increased due to the general scarcity of grain both in Britain and the rest of Europe. This was a matter of immediate concern to those who needed food and who could not even find enough work to earn the cash necessary to buy meal. '*At such rates as we ought to charge.....it is impossible for men to maintain their families above the level of starvation on the ordinary wages of labour*', stated Coffin to Trevelyan. The latter replied: '*We cannot force up the wages of labour, or force down the prices of provisions, without disorganising society.'*

In an effort to persuade land-owners that they could provide much-needed work for their tenants, payment for which would be in foodstuffs, Coffin drew their attention to certain Government Acts which allowed the Treasury to offer State funds for estate improvements. One such Act was the Drainage Act, originally passed to help the destitute Irish, and which now enabled Highland proprietors to use large numbers of crofters as ditchers on their estates. But the response was poor, with some landowners preferring to resort to efforts to relieve suffering from their own resources. Not all proprietors had conscience enough to see to it that deaths from starvation would not be laid at their door. But a number took up the challenge. At one time MacLeod of Dunvegan was feeding, through employment, some 8000 men on the Dunvegan estate; among that number were folk from neighbouring estates whose proprietors did not take such a generous step. On the island of Lewis, the new proprietor, Sir James Matheson, disbursed some of his profits from drug-trading on famine relief. Lord Lovat helped his tenants on his North Morar estate, and MacLean of Ardgour took the unusual step of training his tenants to grow peas, carrots and cabbages as substitutes for potatoes. This activity by MacLean brought the comment that it indicated an unanswerable reply *'to those who are always exclaiming against the inverate indolence and incorrigible obstinacy of the Highlanders.'*

These were only the bright spots in an otherwise darkened sky. The recent purchaser of South Uist, Benbecula and Barra, Gordon of Cluny, showed the other side of the coin. An investigator who visited those islands reported *'an awful reflection ....that at this moment the wealthy heritor of this island is not employing the poor population'* and predicted that *'scenes will occur in South Uist, Barra and Benbecula which would be disgraceful to his name, and injurious to the reputation of Great Britain'.* Reading this Report, Coffin wrote off to Gordon of Cluny and threatened *'to interpose in favour of the sufferers..... leaving to Parliament to decide whether or not you should be legally responsible for the pecuniary consequences of this just and necesary intervention.'* Gordon reacted by starting estate work which was paid for by doling out quantities of meal to alleviate the situation in which his tenants found themselves.

Lord MacDonald of Sleat had some 14,000 tenants on the verge of starvation, of whom some 400 were employed on a drainage scheme which was abandoned after a few weeks. His factor actually purchased a considerable quantity of meal in Liverpool and then decided that he could not transport it to Skye without being *'forced to sell it at a great loss.'* He therefore re-sold the meal, and, taking advantage of the rising prices for grain, he made a handsome profit. And while the profits were being counted, Lord MacDonald's starving tenants had to make a weekly thirty mile trek to Portree to procure minimal supplies of food from the Government depot.

While the problem of mass starvation facing the crofting population was one quandary requiring the full time attention of officials, another was raising its head — the lack of seed potatoes for planting in the spring of 1847. In many areas seed stock was non-existent, for the simple reason that it had been eaten. In many areas seed stock was so low that its planting would hardly yield sufficient to prevent another year of starvation and destitution. The Government was unwilling to supply new seed stock, and tried to persuade the landlords to do this, without much success. The overall result was that good land had to lie fallow.

Other bodies which came upon the scene to help alleviate the problem included the Free Church, which became as active in Catholic areas of the Highlands as it was in the Protestant parts. Most of those charitable bodies eventually became united under a Central Board of Management of the Fund for the Relief of the Destitute Inhabitants of the Highlands, a long-winded title for an organisation which undertook to take overall responsibility for fund raising and the distribution of food from a cash inflow which had reached over £250,000 by the spring of 1847. And all of it was needed. The Board had hoped that its efforts would ensure that 1847 would see the end of the problem. But in the early autumn bad weather flattened crops of corn and a new outbreak of blight devastated the scanty sowings of potatoes. Reports again started to come in about pathetic conditions in all parts of the Highlands. In South Uist:

> 'The scene of wretchedness which we witnessed as we entered on the estate of Col. Gordon was deplorable, nay, heart-rending. On the beach the whole population of the country seemed to be met, gathering the precious cockles.....I never witnessed such countenances — starvation on many faces — the children with their melancholy looks, big looking knees, shrivelled legs, hollow eyes, swollen-like bellies — God help them, I never did witness such wretchedness.'

The Central Board then set up Local Committees to oversee a new phase of relief work, but ended up with a system which consisted of '*a huge staff of stipendiaries on liberal pay, and multitudes of starving supplicants receiving a modicum of meal.*' Not only that, but crofters found themselves being subjected to a means test before they could receive any food. Eventually the system of relief was based on public works, such as the many 'Destitution Roads' which were built by hungry men who were paid low rates for their labour. A contemporary Report said the men saw:

> '.....the injustice of being paid at a very low rate out of what they not unnaturally consider their own money and are exasperated at seeing gentlemen living in comfort on what they know was subscribed to them, while they have to walk, often without shoes and always in insufficient clothing.....to the source of labour, where, after working for eight hours, they receive the value of 1½d.'

Only very slowly did matters begin to improve for the crofting population, many of whom decided to take the long road out of the Highlands by voluntary or enforced emigration. Those who were left behind, or who opted to hang on despite pressures to remove them from the land, were those who some thirty years later were to become active in the agitation for land reform in the Highlands. History had been born in hungry bellies, in the unmarked graves of thousands who died from starvation, and, not least, in the attitude of landowners who had so cruelly denied their moral responsibility for the people who lived on land which they had obtained as a result of their privilege of wealth.

Ferrying Cattle Across The Sound of Vatersay

GOD'S GRACE ON THE CATTLE

It is not often realised that the physical nature of the Highlands and Islands is such that raising of livestock has more potential for a stable economy than has the scraping of earth to grow a few crops. When one journeys round the various parts of the region, only those areas with fertile glens are immediately thought to be suitable for livestock. Yet, particularly during the sixteenth and seventeenth centuries, the region's economy was based on cattle raising to such an extent that vast herds of cattle were drawn from every corner of the mainland and the islands to travel to the cattle trysts at Crieff and Falkirk, where dealers were waiting to trade for meat and hides to satisfy the demands of the English market.

Considering the primitive conditions in which those cattle were raised, it might come as a surprise to learn that by the early years of the 17th century, over 20,000 cattle were sold at those fairs. When the main cattle tryst was shifted from the traditional stance at Crieff to Falkirk, to satisfy the needs and convenience of English drovers, that number increased. It was estimated that by 1777 some 30,000 head of cattle changed owners at the thrice-yearly sales, a figure which rose steadily until about 1850, when a peak of 150,000 cattle were sold at each sale.

Those vast numbers came from all over the Highlands and Islands. The river of cattle first originated in small trickles from the islands and remote mainland townships, to be joined by those from neighbouring districts. These were then met by others until a black flood of beasts made its slow way south. Cattle from the Hebrides were ferried over to Skye in small boats, and from Skye they swam to the Scottish mainland at Glenelg, over the Sound of Sleat. The whole of the Highland economy was once almost wholly based on the cattle trade; on it were dependent laird and clan chief, tacksman, tenant and clansman. And it was when the trade was disturbed by local feuds, civil uprisings or disease such as the cattle plague of 1781, that its real import was felt, for the laird ran into serious debt and the clansmen faced starvation.

The keeping of cattle stock was always an important activity for the Highlander, even if that activity was limited to the tending of a few cows. Without those animals, the whole family might starve; as it was, the cattle provided milk and its by-products, meat, leather and even horn for such simple items as spoons and buttons. Dairy products were often the only source of food over a considerable part of the year, and any crops grown were mainly intended to provide animal fodder rather than food for the family. In the direst of days, when starvation knocked at the door, the cattle were often bled, the blood mixed with oatmeal and cooked to make a poor supper.

In many areas of the Highlands, dairy produce was a kind of currency. Rents were paid in butter and cheese; surplus produce was often sold to buy meal in order to supplement inadequate home-grown supplies or to buy necessities such as iron and fishing hooks.

The herding of cattle was a job for young and old, particularly in those areas where cultivated land was unfenced. While the young crops were growing, the usual practice was to take the cattle to hill pastures, the practice of *'transhumance'* which is also known in Norway and Switzerland. By transferring the herds from the township grazings, in much need of a relief from grazing to get back into form in time for the winter months, the sweet young grass of the slopes could be used to fatten catle at no expense. The practice involved women and children making their homes in rough bothies known as *'shielings'*, leaving the menfolk to look after the township and its needs, such as the repair of houses and their re-thatching.

While at the shielings, the women and girls made milk products such as butter, cream and cheese, which were then sent down at regular intervals to the township for immediate consumption and storage for the winter. Those shieling sites can often be seen today as vivid patches of green among seas of brown and purple heather, and the practice is now no more than a fond romantic memory commemorated in the love songs of last century, born in the mists of Celtic twilight when expatriate Highlanders' memories had erased all traces of the bad times and could only recall the balm of happy summer days.

When the sheep and Lowland flockmasters came to the Highlands, the almost limitless grazings were lost and the crofter found his cattle confined to delineated areas, and was often forced to reduce his livestock to what his new acreage could support, unaided by the lost hill grazings. What grazings were left became the common interest of the township and, in order to regulate them, township constables were appointed. Those officials were responsible for the oversight of livestock and ensuring that they were kept clear of cultivated ground. They also oversaw the '*souming*', that is, the control over the number of animals a crofter was allowed to keep on the common grazings, that number depending on the size of his holding and kind of livestock kept.

The constable had the important job of keeping a careful note of every man's stock, and his management of the common grazings was often a matter of vital precision to prevent over-grazing. A '*soum*' is the amount of pasturage needed for a cow; this varied from district to district throughout the Highlands and Islands, and because a crofter's livestock consisted of more than cows, a variety of equations, using the cow as the common factor, was evolved. Again, this could vary between districts. In general the following equivalents obtained of equating a cow against other stock, and was known as '*coilpeachadh*'. It was in this way that a crofter's preference for cows, sheep, horses, lambs and calves, all in various stages of maturity, could be honoured.

One cow was equal to: ½ horse; 8 calves; 4 stirks; 2 queys; 8 sheep; 16 lambs. One horse was equal to about 16 sheep.

Cattle Landing On Barra After Swimming The Sound Of Vatersay

One interesting aspect of the application of *'coilpeachadh'* to the common grazings held by a township was that, apart from satisfying each crofter's livestock holding, the system provided for the conservation of the grazings, in that each animal's grazing habits were taken into account. A sheep, for instance, will eat out the finer grasses and allow coarse species to grow.

If a crofter could not take up his soum, he could let the difference to a neighbour who might be overstocked, with the agreement and consent of the township constable. When the Napier Commission, set up in 1883, went round the crofting communities seeking evidence about the social and economic conditions of the Highlands and Islands, it received a unique insight into the workings of townships, and was so impressed by the livestock regulating system that it recommended, in 1885, the retention of the township constable, freely elected by the crofters.

In the event, the Act of 1886 did not implement this, and the position was taken over by a Grazings Committee for each township, run by a Clerk who has no executive or disciplinary power. The result has been that over the years, as the working and stocking of a croft has become less a life-and-death subsistence matter than it was, the old souming rules have become neglected. In one respect, this is now seen in the overgrazing by sheep stock pushed to a dangerous extent and the consequent reversion of land to an infertile condition. In addition, the neglect of the old custom of going to hill pastures in late spring and summer, thus relieving the in-bye land of grazing and giving it a chance to recover, has added to the problems of over-grazing.

Only in a few areas does cattle-raising now play an important part in the crofting economy. In Mull, for instance, the cattle stock is as low as 30 percent of the figure of 60 years ago. Indeed, the Report of the Committee on Crofting Conditions in 1954 indicated that in the north-west of Scotland *'.....there is not an average of one cow per croft.'* Another important change has been the distortion of the old ratio of one cow to eight sheep. The ratio has now changed for the worse, to one cow to thirty sheep on average, and in some areas the density is as much as fifty sheep to one cow.

Some parts of the crofting counties, however, have always had a tradition of association with cattle, and the crofters there have maintained a lively interest with good economic returns. One such area is North Uist, where the agricultural economy is based on the rearing of beef cattle, an interesting echo of the situation recorded in the Old Statistical Account for Scotland (1793), where it is mentioned that North Uist supported about 2000 cows, of which some 300 were exported every year. This is still the traditional activity. The cattle are a blend of Highland and island strains which gives a breed called 'Luing'; some of the breed show the shaggy, wide-horned Highland strain more strongly than others. The calves are highly valued and buyers attend the island sales to buy these animals, which are often in peak condition after having spent a summer on the island's summer grazings.

In the old days of the droving industry, the cattle were known as 'Kyloes' or black cattle, from the predominant colour, although duns, browns and brandered hues were also common. They came to the notice of some of the agricultural reformers of the late eighteenth century who admired the Highlanders' *'hardy, industrious and excellent breed of cattle.'* They were said to be *'short in the legs, round in the body, straight in the back'*, but small in size. In time, black was bred out in favour of the brown and dun, and estate owners delighted in having them around the homestead as walking ornaments — to the detriment, incidentally, of the breed, which tended to be bred for length of horn, rather than for the traditional virtues of hardiness and ability to survive in bad conditions, and to rear a hardy calf in those bad conditions. Other breeds also became popular, such as Shorthorns and Aberdeen Angus. But in recent years the old Highlander has come back on the scene, and, with some careful breeding and cross-breeding, the old familiar cattle of a couple of centuries ago are again making a major contribution to the meat market.

There is still a great potential in the crofting counties for cattle raising, a potential slowly being realised. One day crofters will learn from history, and become major producers of beef again.

BLUE-SMOKED PEAT

In the popular mind, peat smoke symbolises crofting. Even in recent escapist literature aimed at the Highland tourist, the deliberate mention of the reek of peat smoke as one approaches a crofting township reflects this peripheral aspect of crofting, as though it were the primary function of the crofter to keep his peat fires burning on his economic home front, and especially so if he relies on the tourist to provide part of his annual income. But it is not so difficult to appreciate how peat smoke has become synonymous with crofting. For some two centuries now, books written about the Highlands and Islands have emphasised the peat fire as the centre of the domestic and communal social life in old Highland villages. Round the fires were told the stories of the ancient past, were related the events of the recent past and the present, were sung the songs of heroic deeds and love. But the peat fire was seen in other aspects:-

> 'I have been in several of their houses, where I saw aged people tortured with rheumatism, sitting amidst thick volumes of blinding peat reek, the only window to the apartment being an open aperture in the wall without any glass, the wind blowing in and whirling the reek hither and thither, and out at the door.'

That was said by Malcolm Ferguson, a Gaelic speaker who visited Skye in 1882 and had little sympathy with the crofters from Glendale who, in March 1883, were sentenced to two months' imprisonment in Edinburgh for agitating for action on the crofters' rights issue.

Bringing Home The Peats, Lewis

Another traveller, William Macgillivray, was moved to write:-

> 'In the meantime the natives were snugly seated around their blazing peat-fires, amusing themselves with the tales and songs of other years, and enjoying the domestic harmony which no other people can enjoy with less interruption than the Hebridean Celts.'

A more reasoned picture was painted by a Dr. James Johnson in 1834 when he was forced to seek shelter

> '.....in a wretched-looking hut, built of rude stones and thatched with heather. We found the interior much more comfortable than we expected. A good peat fire was blazing on the hearth, over which was suspended a pot of broth; while around the chimney hung more than a dozen well-smoked salmon and other fish.'

Peat in the crofting areas has been a long time in the making, and is still a growing product of the Highlands landscape. In what is called the *'sub-Boreal period'* from 4000 B.C. to 500 B.C., the climate was sufficiently cool and dry to allow the growth of such trees as oak and elm. By this time Man had arrived in the Highlands, to erect the Callanish Stones in the Bronze Age, some 3000 years ago — a monument which has been invaluable in dating the Lewis peat to a period subsequent to their erection. In 1856, Sir James Matheson, proprietor of the Lews, excavated this veritable Stonehenge and cleared away some five feet of peat growth. Just before the Christian era there was a change of climate: still cool, but now wetter. Those conditions were ideal for the development of blanket bog, and those areas based on the hard Lewisian gneiss and Torridian rock succumbed. The conditions for the growth of blanket bog included an abundant and fairly evenly distributed annual rainfall, cloudiness and a temperate climate of cool summers and mild winters, all of which are typical of the north and north-west of Scotland. The result is seen today: hundreds of thousands of acres of barren land which, in a few places only, are being tickled back into agricultural use by reclamation.

Peat is found in several varieties. It may range from a soft, plastic, yellow-brown, amorphous substance, to fibrous peat which is brown in colour and frequently contains scarcely-altered plant remains, and even roots and tree trunks from a long-by age. In many places where the water table is very high, the peat may be formed from sedges and rushes. Peat bogs also vary in thickness, in some cases only a few inches, while in other localities it has been found to be over 12 feet thick.

When the land of the Highlands was better managed, before the decline of the clan system and the onset of the era of change heralded by the Jacobite Rising, much of the peat lands supported the Highland population. At the present time land use in the region includes forestry, crofting, mixed and arable farming, deer forests, grouse moors and pleasure (the last usually of the few to the exclusion of the many). The present patterns of land use stem from a history which is varied and paradoxical, and which contains the elements of today's social feelings.

Up to the 17th century, the land was held in trust by the clan, or in individual peasant holdings. During the 17th century, royal gifts of large tracts of land were made to influential individuals who cleared much of the existing forest cover and allowed heather and bracken to gain a hold which is still as tenacious as ever. By the end of the 18th century, much of the land was in the hands of a few individuals, and smallholders found themselves being cleared out in enforced emigration schemes. Between 1763 and 1775 no fewer than 20,000 took ship to settle in America, and most of them were from the north-west Highlands and the Islands. Highland black cattle, a former economic mainstay, disappeared, their place being taken by sheep which rapidly caused formerly well-tended grazing to deteriorate. A coarser vegetation took over. To improve the situation, heavy burning operations were undertaken by graziers, but in certain types of soil this had an adverse effect: the destruction of soil nutrients. Then, diminishing returns from sheep farms forced the introduction of the deer forest and deer stalking as an economic activity. The vacuum left by the departure of the sheep was filled with deer, and the deer forest area of the Highlands increased tenfold in a period of fifty years. Next came the Forestry Commission, which, with its planting, takes vast areas of land out of use for half a century at a time, a period during which the land becomes infertile and presents an insurmountable economic problem to put back into agricultural use again.

Faced with a land history like that, it is little wonder that the crofter has displayed little incentive to put his accessible peatlands to good use. And yet, in recent years, this is being done by draining, treating with shell sand and lime, and then sowing seed to produce a good sward for grazing.

But for the most part, the peat bog is simply a source of fuel for the crofter and has ever been a godsend in areas where there is little or no woodland or scrub cover. Without the peat there would have been nothing to burn and the folk would have cleared themselves out of the Highlands and Islands, for without fires there can be no domestic or social life.

The peat smoke might have offered another kind of blessing. In 1885 the Inspector for Health for the North Highland District wrote:-

> 'It is not surprising that, under such conditions, typhoid and typhus fevers and other forms of disease usually attributed to bad sanitary arrangements, should be of frequent occurence; it is more surprising that any person should escape. The only explanation I can suggest is, that possibly the poisons arising from so many forms of pollution, within and without the houses to which I refer are counter-acted by the constant burning of the open fires usually placed in the centre of the houses, and by the abundance of mountain and sea air which is admitted by the open and ill-fitting doors. It is also not impossible that the dense clouds of peat smoke in which the people continually live may have some salutary antiseptic properties.'

For many crofters, peat-cutting is an important period in the year, though it is not now, perhaps, the springtime social occasion it once was. In former times, the township shared in the cutting of the peats for all the families, and particularly for folk who were either aged or infirm. Nowadays, it tends to be a limited family occasion. Dried peat is a good clean fuel and generally has about one-third the calorific value of coal. But its winning takes an immense amount of labour. A good peat cutter will cut up to 1000 peats a day if he has adequate help to lift the wet turves and lay them around the cut bank. Peat contains over ninety percent of water, and so the freshly cut slabs must be left to dry sufficiently until they become firm enough to handle for the subsequent air-drying. Each peat, after cutting, is handled at least twice before being taken back to the croft and stacked.

The peat is cut with a special but simple tool called the *'tosg'*, *'tairsgeir'* or *'toirsgian'* depending on the area or locality, and is made up of a long handle with an angled blade, the detailed shape of which also depends on locality. The implement cuts on two sides at right angles and allows the cut peat to slide up the shaft, which is flattened and slightly broadened above the cutting blade. The soggy peat is then lifted and thrown as far away as convenient, so that, once a bank has been cut to two or sometimes three peats' depth, the immediate area round about is covered with turves, ready for a dry spell of weather which it is hoped might last for a week or more. Thereafter the crofter returns to gather the peats and form them into small stacks. Drying patterns are legion. In some areas a wall is built of overlapping peats, with air gaps between each peat. Again, four peats are made to stand edge to edge to form four sides, with two further peats balanced on top. Or else two peats are propped against each other, with a third on top.

Once the peats are all dried hard, they are taken to the nearest access to a main or moor road where they are loaded onto a lorry or tractor and carted to the croft, there to be built into the familiar peat-stacks which are often part of the landscape of the crofting township. Many stacks built by the older generation are works of art in themselves, their general shape reflecting the older design of black houses. The outer layer of peats protects those inside the stack from rain and bad weather because once a peat is thoroughly dried out it never absorbs its original water content. A household fully dependent on peat for fuel will require about 15,000 to 18,000 peats each year, which represents nearly 20 full man-days work each year, not counting the work of others in throwing, initial stacking for drying, taking to the roadside (nowadays in bags: no longer the old home-made creel), and then the construction of the final stack. With the rise in the cost of coal in recent years, many households in the rural crofting areas have gone back to being more dependent on peat as a main fuel, particularly now that the design of solid-fuel burners has become more sophisticated to provide facilities for cooking, hot water and central heating.

The best type of peat is the non-fibrous variety, with the colour of dark chocolate; it burns slowly with very little flame but emits great heat and leaves a fine, dust-like ash.

It is not surprising that peat has entered into the Gaelic language and its traditions in many ways which reflect the close association of the croft with its necessary fuel. For example, there are many names used to describe parts of a peat-bank, a litany which often reflects variations in particular areas. Even in Caithness, not a prominent Gaelic-speaking area, there are names, now falling out of use, to describe those peats cut with an ordinary spade and not the traditional peat-cutter.

Peat fires, once lit, were never put out, although in an older time it was the custom, on Beltane Eve, to douse all fires save one in a township, a lighted peat from that fire then being used to re-light all the others.

At the close of each day, the fire was '*smoored*', that is, given a dousing of small peats and dust which would smoulder away during the hours of sleep until it was fanned into flame the following morning. There was a very practical reason for this care, because kindling was scarce, particularly in the islands; it was considered a major domestic crisis if the woman of the house was neglectful of her duties to keep the fire alight.

That the smooring of the hearth was an act of significance to the household is reflected in a beautiful prayer, collected by Alexander Carmichael last century:

I smoor the fire this night
As the Son of Mary would smoor it;
The compassing of God be on the fire,
The compassing of God on all the household.
Be God's compassing about ourselves,
Be God's compassing about us all,
Be God's compassing upon the flock,
Be God's compassing upon the hearth.
Who keeps the watch this night?
Who but the Christ of the poor,
The bright and gentle Brigit of the kine,
The bright and gentle Mary of the ringlets.
Whole be house and herd,
Whole be son and daughter,
Whole be wife and man,
Whole the household all.

Finally, a description of peat-cutting in South Uist, 1890, from the excellent and evocative writing of F.G. Rea, an English-speaking teacher, who found himself schoolmaster in a Gaelic-speaking community:-

> 'Saturday morning came with the sound of men's voices, the clattering of feet, and ring of metal on stones near the house. My party of peat-cutters had assembled, fourteen of them all told, with Sandy at their head. I led the way into school where a long table improvised from school benches had been set for breakfast. They followed me, leaving their tools outside. My sister and Catriona were busy preparing fried ham and eggs, and strong tea in large pots and jugs. Piles of scones, butter, cheese and jam were already on the table at which the men seated themselves. A hot plate of ham and eggs with a cup of tea was placed before each and to my surprise they all sat solemnly still gazing in front of them without saying a word.
>
> 'I was wondering if they were waiting for me to say

'grace' when Catriona whispered in my ear: "The whisky!" This was produced and white-haired Sandy passed round the table handing each a brimming wine-glass of whisky, the contents of which were tossed off without a blink or word and with the utmost solemnity. All sat silent till the last man had drained his glass, then first looking round at each other they began to eat heartily of all before them. Before long they were talking to each other in Gaelic and seemed more natural — the effect of the food and perhaps the whisky, thought I.

'When they had breakfasted to repletion — and I noticed each man said a grace of thanksgiving, signing himself with a cross as he did so — an ounce of tobacco was given to each from which they shaved and rolled a pipeful, then they sat comfortably smoking and talking in low tones. At a word spoken in Gaelic by Sandy, they all rose and, putting on their caps at the school door, took up their tools and started for the peat banks.

'I was curious to see this process of peat-cutting, so followed. We arrived on a raised piece of grass-covered ground which was intersected here and there with ditches. Four or five of the men threw off their coats and with spades removed the turf. A smooth damp-looking close-fibred brown substance was then revealed. The white-haired leader told me that this first process was known as "*skinning the banks*" and the ground was now ready for cutting the peat. Two men now came forward to the front edge of the raised peat bank and one of them jumped into the ditch below. The other man carried in his hand a most strange implement called "*the iron*". The upper part was just like that of an ordinary spade, but the lower part, made of heavy steel about fifteen inches long and three inches wide, had a ten-inch keen-looking knife-blade some eight or nine inches from its end. This man advanced to the front of the bank and placed the base of the iron on the peat so that the knife-blade was directly to his front with its point exactly on the front edge of the bank. He then pressed a foot on a protuberance on the shank of his implement just above the blade and the iron part sank into the damp peat to its own depth. With a jerk of the handle a clean-cut piece of peat fell from the bank into the hands of the man waiting in the ditch below who immediately threw it with a sliding motion of the hands on to the bankof the other side of the ditch. Stepping back the length of the knife-blade, the man on the bank cut another piece, and so on for the whole length of the peat bank, one man cutting and jerking, the other catching and throwing with such regularity that the movements appeared to be automatic.

'About six o'clock word came for me at the house that I was wanted at the peat banks. I found the crew standing resting on their spades and their cutting implements. Sandy said they wanted me to inspect the peat they had cut and to know if I were satisfied. The ground was strewn with cut peat in all directions; they had cut three "iron" depths of peat from each bank, about a hundred cart-loads I was told. Of course, I thanked them and led the way to the school for tea. I am sure they well deserved the hearty meal which they ate and the extra dram and the twist of tobacco which I gave them with a shake of the hand on parting.

'After the clearing away and the washing-up of crocks, glasses and the knives, forks and spoons, my sister, mother and I experienced a mutual feeling of thankfulness that all was over for the day; and as we stood together in the garden that peaceful summer evening, drinking in the soft air and the calm beauty reigning around us over land, hill sea and sky, I am sure we thanked God in our hearts for all that had been vouchsafed us.'

Wash Tubs and Washing Pots, Loch an Duin, Lewis

WOMEN'S WORK

Highland women have seldom been given their proper place in the history of their country. Flora MacDonald might easily spring to mind as an example of their bravery and courage in the face of that peculiar set of circumstances which threw her into the public eye and retained her memory in the pages of popular and palatable Highland history. Yet, in many ways, the greatest burden of historical events has always fallen on women, who have had to bear the brunt of tragedy, wreaked by nature and man-made. Far from being only the keepers of the home and hearth, they have also had to be the caretakers of the traditions, the morals and the intrinsic values of their communities, to say nothing of the occasions, particularly during the troubled times of last century, when they took their stance against the forces of law and order to make a forceful point for their rights and for justice.

This in fact was the situation in which women found themselves during the hard years in the nineteenth century when faced with eviction from their homes, often when their husbands were conveniently away from the distict, fighting for Queen and country in such war arenas as the Crimea. One such incident was their opposition to the law and order forces in Strathcarron, when they refused to accept writs of removal intended to be served on them and their families. A low-level skirmish had already taken place during which the summonses were taken from the Sheriff-Officer and torn up. The Officer then returned to report his failure to carry out his duties. So:

> '......private meetings were held in Tain, the great heads of the evicting firms and the great sheep lords consulted together, and it was resolved upon to go into the district with a strong police force in order, as they said, to uphold the majesty of the Law, and to strike terror if not into the hearts at any rate into the skulls of the offending females.'

So it was that on Friday March 31, 1854, the eviction squad, a muster of more than 40 men, marched into Strathcarron. Four miles down the glen they met some 60 women, with a dozen or so men standing behind them. The women waited silently. The Fiscal shouted to them to get out of the way, and when he got no response, proceeded to read the Riot Act, although this was disputed later. When there was still no response from the women to allow the police to proceed on their way, he gave orders for the women to be charged. The police went forward with truncheons drawn, at which some of the men of Strathcarron fled, to leave their womenfolk to meet the approaching line of batons. The women stood their ground with determined faces, armed with a few stones. In the ensuing melee, only a few policemen received the odd bruise or lost a helmet. For their part the police:

> '.....struck with all their force.....not only when knocking down, but after the females were on the ground. They beat and kicked them while lying weltering in their blood. Such was the brutality with which this tragedy was carried through, that more than twenty females were carried off the field in blankets and litters, and the appearance they presented, with their heads cut and bruised, their limbs mangled and their clothes clotted with blood, was such as would horrify any savage.'

So many women had sustained such severe injuries that they carried the marks of the encounter for years afterwards. So much blood was, in fact, lost, that the ground had to be ploughed up to hide the evidence of the brutality wreaked by the forces of law and order.

Women were also in the forefront of another encounter with law and order at the Battle of the Braes, in Skye, in April 1882, where the women, '*by far the most troublesome assailants*', opposed the summonses which would have evicted them and their families from their homes. The incident was the turning point in the crofters' war against landlordism and the biased forces of law and order arranged against a people who had no legal rights, let alone standing. The Braes episode advertised the crofters' cause and revived memories of the atrocities committed in the days of the clearances and the enforced emigrations. From the fractured heads and limbs of the Braes women came the Napier Commission which was to pave the way for security of tenure for crofters.

By and large, the role of women has been thirled to the domestic scene, to the dairy work, and in the production of cloth. But they also performed many laborious tasks outside the home environment. In 1823, Hugh Millar wrote that in the Gairloch district, the men dug the land with the '*cas chrom*' and sowed it while '*the wife conveys the manure to it in a square creel with a slip bottom, tends the corn, hoes the potatoes, digs them up, and carries the whole home on her back. When bearing the creel she is also engaged in spinning with the distaff and spindle.*'

These were, however, only a few of the many domestic chores to which the women had to turn their hands, both in hours of daylight and darkness. The milking, for instance, was a daily task, the product afterwards having to be stored in shallow basins before being skimmed of the cream, using scallop shells with holes pierced in them to let the thin milk drain away. The wearisome task of butter making also fell to the women, as did the making of cheese. These products were of vital importance to the family's use, but as much surplus as could be spared went to pay the rent; they were also among the few products which could be sold for cash to pay for items needed in the home and on the croft, and which could not be home-made. In the eighteenth century, those dairy products were used instead of currency; it is not often realised that a cash economy was not introduced into the Highlands until late in the nineteenth century.

The skills in wool dyeing, carding, spinning and weaving which

Instruction On The Spinning Wheel, c.1910

were developed over the centuries for the making of cloth for domestic use or for barter stood the crofting women in good stead in the latter years of the nineteenth century, with the beginnings of the home-made craft trade being found in the aftermath of the potato famine of 1846 and 1847.

Towards the end of the last century there was an increasing interest in home-craft products, mainly fostered by the upper classes. In 1889 the Scottish Home Industries Association was formed by the Countess of Roseberry, who had the foresight to see that unless the products of the croft were exposed to organised marketing methods, the producer would have little chance of making a significant contribution to the household economy, to say nothing of obtaining a fair price for the goods.

As the scheme progressed and markets were developed for a wide range of home-craft goods, it became evident that the income earned in this way was the salvation of many a family in the Highlands and Islands. The Scottish Home Industries Association, after its formal registration as a Limited Company with a nominal capital of £10,000, with depots in London, Harris, Lewis, Golspie and Edinburgh, included in its activities training to improve the quality of produce and offered capital to buy raw materials. Over the years, the Association's work was added to by the efforts of other organisations such as the Harris Tweed Association, Highland Home Industries and the Crofters' Agency, which, in concert, eventually created a craft industry out of what had been home-based activities in a subsistence economy. In many cases, it was the efforts of organisations such as these which first introduced a money economy, and thus had a most profound effect on the whole society and economic structure of the area.

The old skills of knitting and weaving are still very much a part of the crofting economy, with both men and women participating in what is a reasonably lucrative business in sales to tourists and to wider markets. In Lewis, in particular, weaving Harris Tweed is often a major activity by weaver-crofters, whose product contributes not only a significant element of stability in the economy of the island, but also to the maintainence of the population.

THE FOOD WE EAT

Despite the comparative infertility of much of the land of the Highlands and Islands, the larder it was able to provide, supplemented by the generosity of the salt seas and the waters of fresh lochs and rivers, was always sufficient to maintain a healthy diet. It was only in the century of the Clearances when starvation became a familiar deathly figure descending on poor landless families with their fate, that the old days of plenty to eat became almost a forgotten dream. The region in fact was able to provide in plenty for those who were willing to fish, hunt, and till the land. Starvation came only when the folk were driven from the land which had supported them and their forebears so well for so long. The soil, though of diverse quality, and in a cold and wet climate, still offered the opportunity to grow food to a high quality, as well as quantity, so that the houses of the old Highland chieftains were often noted for their groaning tables, open to all and sundry. Many of the travellers who safaried to the Highlands and Islands during the 17th and 18th centuries make frequent mention of the hospitality of their various hosts and wrote of the abundance of milk, fresh cream, cheese, game, poultry, fresh and salt-water fish, mutton and beef. A culinary delight for the Frenchman Fujas de St. Fond was a '*sort of pap of oatmeal and water, in eating which each spoonful is plunged into a bowl of cream.*' He was describing no more than the usual porridge and the Highlanders' way of eating it. Dr. Samuel Johnson, however, was more critical in his observations:

> 'Their tables were very plentiful, but a very nice man would not be pampered. As they have no meat but as they kill it, they are obliged to dine upon the same flesh while it lasts. They kill a sheep and set mutton roasted and boiled on the same table together. They have fish both of the sea and the brooks, but they barely conceive that it requires any sauce. Barley broth is a constantly well made dish.'

He added that he advised any guest to make a point of having some, for he might not be able to eat anything else. Generally, white bread was a great luxury and served only to distinguished guests — the Highlander preferring, very rightly, his oatcakes and bannocks.

It was only when the relations between clansman and chief began to break down that the clansman, increasingly becoming a landless entity and more of a squatter on land now legally belonging to his erstwhile chief, found that his family's standard of living and diet began to deteriorate. People perforce had to learn to live austerely. On the Black Isle, for instance, despite its relative fertility, Sir John Sinclair observed at the beginning of the 18th century that the people lived on meal, dairy produce and sometimes fish, but only in such quantities as to enable them to subsist but '*hardly in a manner adequate to give spirit or strength for labour.*' He quoted typical menus. For breakfast there was gruel and bread, or porridge and milk, or flummery and milk. Dinner consisted of potatoes and milk, bread and milk or sowens and milk. Supper was invariably potatoes, gruel or kail. Martin Martin speaks of the ordinary diet of the Skye folk as being butter, cheese, milk, coleworts, potatoes and '*brochan*' (oatmeal and water boiled together). In 1846 the parish minister in Badenoch wrote:

> 'Potatoes and milk may be said to constitute the principal foods of the peasantry. For the meal which the small tenants can raise on their farms in their mode of cultivating them would not support their families during one third of the year.'

The advent of the potato as a crop in the Highlands and Islands was both a blessing and a curse. Once it had been proved to be a useful food, reasonably easy to grow, the potato became the staple food of the common folk, ousting oatmeal to second place, and was used in large quantities at every meal. After the repeal of the iniquitous Salt Tax in 1817, fish could be salted and stored for winter use; along with the potato, herrings became the principal articles of food. But this increasing dependence on too few items proved to be the downfall of those who neglected the traditional crops, for, in the 1840's, when the potato crop failed in successive years, the spectre of starvation invaded many crofts, and, as in Ireland, took its heavy toll of life.

Even so, the humble diet of the crofter was often ample enough to sustain strength and life, and, in the case of the cautious crofter who did not place too much reliance on the ubiquitious potato, it was possible to have a widely varied selection of foods to choose from, prepared in an astonishing number of ways, many of which still survive today.

Meat came from deer (when the laird's gamekeepers were lax in their duties), sheep and cattle. The cattle, however, were valuable assets on a croft, and were killed only in dire circumstances, for from the cow came both milk and all its products such as cream and cheese. When a cow *was* killed, all of the animal was put to some domestic use: from shoes made from the cured hide, to horn spoons, fat for candles, and even the blood was added to oatmeal and suet to make black puddings. In Badenoch the stick used to prepare the popular frothed milk consisted of a strand of twisted hairs from a cow's tail, fastened round cross-pieces at the end of a thin stick.

Oatmeal was the basis for many kinds of dishes and was, along with barley, used for bread and cakes and bannocks. Meal was ideal for thickening fish soups and meat broth. Oatcakes, now regarded as a special kind of biscuit, were at one time a common enough item in the house, baked on one of the oldest cooking utensils in the world, the girdle (Gaelic: *greadeal),* which evolved from hot flat stones to the now rarely-used flat iron disc, still an excellent method of baking pancakes and other thin breads. *'S mairg a ni tarcuis air biadh'* — foolish is he who despises food — runs an old Gaelic proverb, and it makes sense if one has ever tasted scones homebaked in the old ways.

Fish-liver bannocks were a particular delicacy in coastal areas. Fishermen often left their village shore with plenty of boat bannocks, *'bonnach-eitheir'*, made from oatmeal, butter, minced cod or ling livers, ground cod roe and eggs. It was full of the best kind of nourishment and served the fishermen well on their often days'-long fishing trips.

Fish, of course, was a staple food, derived from many varieties and in particular herring, cod, ling, haddock, halibut and turbot. Fish livers were used in a variety of ways and were prized for their high oil content (no-one then knew of their equally high vitamin content). The fish were salted and sun-dried, or alternatively were rock-dried, that is, left in the wind but out of the sun and regularly splashed with sea water. Thus was food stored for the winter, when fishing was often impossible.

It was with this knowledge of food and its value that most of the folk in the old days managed to maintain a satisfying diet from the simple elements which nature made available from the sea and soil. And even in those areas where the land could not provide large quantities of food, the sea's harvests made up for the land's deficencies. It is not surprising, then, that men and women, nourished on a diet of fish and simple meal, were capable of enduring great demands on their energies in, for instance, the days of virtual slavery when kelp was king.

A Skye grace said before meals indicates the attitude of the crofter folk who, while realising that their own efforts to provide their food might be sufficient, the Divine factor was ever present: *'Bless the table and fill the stomach. Thou who didst send us this meal, may Thou send us another, from meal to meal, until the final meal.'*

Food appears in many old Gaelic sayings. *'Bruicheadh an Leodhasach air a bhonnach — a suidhe air'.* (The Lewisman's way of cooking the bannock — sitting on it.) This is a Uist saying, and is based on the fact that when Lewis fishermen were away for part of the day, they used to take a bannock consisting of two oatcakes with a thick layer of cod liver between them. To cook it, they sat on it during the trip.

'Ceapaire Saileach — aran tiugh, im nas tiughe agus caise nas tiughe na sin.' That is: a Kintail sandwich — thick bread, thicker butter and cheese thicker than that.

'Cha bhi chuimhn air an aran ach a reir fad an sgornain'. Bread is remembered only as long as it stays in the gullet, that is, while it is being eaten.

THE SEA'S ACRES

For those crofting communities located within earshot of the sea, and within easy access of it, the harvest which was obtained from inshore waters was not only an important change in the diet; more often than not it was crucial in the alleviation of starvation when the crops failed. Often, too, when the land required an immense effort to gain little, dependence on the sea as a more amenable option was the main occupation of the crofter-fisherman.

Communities often reflected this dual interest:

> 'Everything is fish; the very little patches of cultivation, recovered at great labour from the rocks and hills, belong to the fishermen, and you often see a man in the fields throwing his sail, with the yard and mast attached, over a stack of corn to preserve it from the weather. When the stacks are thatched and completed, too, the universal covering is an old herring net, bound neatly round them.'

The herring was the prime fish of the sea, and was highly regarded not only for its high protein content but simply because it was caught without too much physical toil. Pickled in brine, it was the main winter food, along with the ubiquitous potato: a bad fishing season was, in some communities, worse than a crop failure. For instance in 1835 only six barrels of herring were caught in Loch Fyne and many families in the area were brought to near-starvation. In 1852 the introduction of the anti-trawling legislation prevented the Tarbert fishermen from pursuing their grade and this, coupled with the failure of their potato crops, produced the situation in which they were reported to be in *'absolute starvation.'*

Collecting Mussels For Bait

Fishing Boats At Castlebay, Barra, c. 1890

Those communities which lacked suitable fishing craft for herring often possessed a boat or two which fished just off-shore for a variety of white fish such as haddock, ling, cod and saithe, all of which could be sun-dried and salted for use during the winter months, a common situation which existed until two or three decades ago. In its heyday, the herring industry offered the crofter an ancillary occupation and a means to earn cash to pay for those items which were outwith the usual barter trade. A Report of 1938 says:

> 'Many types of fishing are practised, but the fishermen in these (the crofting) areas, are predominantly what might be termed crofter-fishermen, that is, men who combine intermittent fishing with crofting. There is, however, an important interest in the herring fishing industry, owing to the fact that the Scottish summer fishing takes place largely in northern and Hebridean waters. This has resulted in the salt curing trade and the exporting side of the industry being largely carried on in the crofting areas and a considerable proportion of the labour, mostly female, associated with this branch of the industry is recruited from the crofting counties, principally from Lewis, Barra and Zetland.'

Not all crofting communities had an interest in economic fishing. For example, on the island of Eigg there are no natural harbours or creeks where boats could lie safely. The crofters on South Uist, whose townships are on the western seaboard, rarely took up fishing seriously, not only because their coast is mostly shell sand beaches which provide no shelter, but also because they directly face the deep and stormy Atlantic. Only where there were sufficient havens to develop did the occupation of fishermen and part-time crofter take on an economic importance. Thus, on Loch Roag, on the west coast of Lewis, there was a fleet of sailing boats which caught ling. The fish were split and dried for export. This thriving community suffered eventually from the advent of the steam trawlers which swept the fishing banks within reach of the Loch Roag boats, and added insult to injury by

Stornoway Pier, 1906

Fishergirls Gutting And Packing Herring, Castlebay, c.1928

destroying the boats' gear. Now, as those Lewis fishermen were always short of capital, it often proved beyond their ability to lay out more cash to replace their gear. In the end, their boats had to be beached and the men went off to the First World War. After that traumatic experience, those who survived the holocaust returned to find their boats rotten and no longer seaworthy. The Loch Roag ling fishery was never revived.

Various reports issued by Government bodies over the years have told the sorry story of the decline in the number of crofter-fishermen, in the Hebrides and from Cape Wrath down to Mull and the Small Isles, which is also the story of a significant change in the economic life of those areas. Only in crofting areas where the land is very poor has an attempt been made to march with the changing times to develop the fisheries. Such a community is on the island of Scalpay, in Harris, where the economic base is founded more on its fishing activities than on its crofting. *'If the fishing's good, there will be a wedding every week in the winter on Scalpay',* runs a saying which reflects that island's dependence on fishing. Loch Fyne is another community where the folk have always looked more to the sea than to the land for their livelihood.

Up until the 1850's, the crofter-fisherman caught his fish from the rocks, or else ventured out a little way offshore in small and frail craft to take advantage of the species of white fish which could not be caught from the rocks. In that way, a monotonous diet was broken during the spring and summer months, and a store of preserved food built up for the winter. The fish also provided such valuable items as fish-oil for lighting, and were appreciated as being health-giving, even though the very idea of trace elements, minerals and vitamins was unheard of.

In the early days, and particularly where no boats were available for sea fishing, estuaries were often dammed by loose-stone dykes, so that fish, coming in at high tide,were left stranded when the tide ebbed, and so were easily caught. Another method used was for bait to be thrown into the sea above a previously lowered net, which was then raised slowly as the fish gathered for their free lunch. When a boat was available, a venture out to sea, a mile or so from

Fishergirls, Lewis, c. 1925

shore, was profitable, when small lines, about 20 fathoms long, were used, with hooks baited with lugworm, sand-eels, limpets and mussels. Collecting the bait was often the task of the children, who had to spend long hours on a cold shore-line gathering what they could for their fathers' lines. Often the catch from those small lines, mostly haddock and whiting, was itself used to bait the long or great lines. These were often left in place for up to 48 hours before being hauled. The long lines were kept in place by buoys whose positions were carefully noted by using landmarks. The catch of white fish was used partly for domestic purposes and the rest sun or wind dried, to be sent away to the Scottish mainland for cash, or else was used to re-pay the debts which most crofter-fishermen regarded as an ongoing part of life.

Each great line had up to 1000 hooks, each hook tied to the line by a snood, often made from horsehair. The hooks were baited by the womenfolk, a long and tiring job. As the hooks were baited, the line and its hooks were skilfully arranged on a flat basket or tray so that it could be paid out without tangling as the boat was rowed along. The buoys to which the lines were attached were originally made from the skins of sheep or dogs, and latterly of metal or glass.

It was those long lines which caught the economic fish: cod, ling, skate and turbot, for which there was always a constant demand. Cod and ling were often gutted before being handed to the curers, and the heads, livers and the three-quarter section of backbone retained. Those items were regarded as belonging by right to the fishermen, along with such fish as skate and dogfish which were neither exportable nor commercial. The dogfish, in particular, was a valuable bonus, for its liver produced oil for lamps,and the fish was also dried for winter use.

Pickling the fish was a long drawn-out job. The fish were placed in large wooden vats containing brine, after which they were laid out on stones where the air could dry them off. When completely dry they were then stacked and baled ready for selling in the markets of the south, such as Greenock, Glasgow and Liverpool.

It was the advent of the herring fishing industry which provided

Fishergirls, Lewis

the opportunity for the crofters to participate in the wealth which this activity created, in that most members of the family could earn much-needed cash by working at the various herring stations set up all round the north and western coasts of Scotland. Most of the boats were, however, the property of lairds, fish-curers or local merchants, who employed the fishermen for a wage. In an official report of 1849, it was stated that along the 50 miles of the Caithness coast, for instance, there were no fewer than 1500 open, undecked boats, manned by upwards of 7000 crewmen. The years up to the First World War were often bonanza times for the industry, with fortunes being made and lost by fish-curers and boat owners. In general, however, the crofter-fisherman, as a wage-earner, or on bounty pay, did not glean great riches from his work. Often during the herring season he had to go into debt with his employer's stores, when the prices he was charged were higher than if paid for by cash. At the end of a bad season, a certain amount of debt often remained to be paid off, and this enabled the employer to hold the fisherman in thrall. Often the fisherman would see out the season by working as a hand on east coast boats which followed the passage of huge shoals of herring round the British coast as far south as Yarmouth.

In these present times when so much is being talked about the conservation of fish stocks, and often so little put into practice (excepting the recent ban on herring fishing in the Minch), it is often difficult to imagine the quantities of fish caught in the boom years of the industry. A report in a local newspaper described the scene in Stornoway in 1924, when over 200 herring drifters worked from the port:

> 'The myriad paraffin flares lighting up the whole quay from Bayhead to the Slip, with hundreds of deft-fingered workers bending over the herring-filled farlings, working with machine-like precision and speed at their somewhat gory task, and men and women in the yellow flare flitting about among the herring barrels, all eager and intent on their work, is more than mere picturesque to the people of Stornoway. To us it is a gladsome

sight, speaking of returning prosperity — after many lean years — in our staple industry. The great hauls landed since the shoals were struck have eaten up huge quantities of curing stock, so that even the best prepared firms found themselves running short of barrels and salt by last weekend. The catch of the last fortnight alone accounted for close on 60,000 barrels, and on Saturday morning, as the boats were arriving heavily laden, it was feared that for want of stock the bulk of the catch would have to lie in the holds till Monday. Fortunately, however, this proved unnecessary, for a stock boat with a large cargo, came into the harbour just as the sales were commencing for the day. It was like the arrival of a relieving column to a beleagured city.'

That description was also typical of places like Lochboisdale, Castlebay, Lochinver, Tarbert and many other towns like Wick, Lerwick and Kirkwall. Tens of thousands of people found work during the season, in trades which have long since died out: sailmakers, ropespinners, coopers, cartwrights, shipmasters, net barkers, sawyers, fish smokers, gutters and packers. And not only men were involved. Fishergirls, in particular, were the mainstay of the industry, the essential link between the fishermen and the final herring-packed barrels carried off by ships from Germany, Scandinavia, Latvia, Poland, Russia and the Baltic ports.

The herring girls often grew up to become gutters and packers of herring and continued at this work until they married. The income they brought into the household was often the means whereby the family could purchase the necessities, and even a few simple luxuries, of life. The girls followed the drifters to Wick, Lerwick, Fraserburgh, Yarmouth, Eyemouth and Lowestoft. The work was hard and the pay was poor, yet to '*follow the herring*' was a way of life for many.

Many girls were engaged on payment of a sum of '*earnest money*' known as '*Arles*', the acceptance of which signified a binding, unwritten contract on both sides, employer's and employee's, for the season's fishing. Working hours depended on the day's catch, but could often be from 6 am until 10 pm. The girls would be wakened by a cooper banging on the door of the hut where the girls lived and slept, and calling: *'Get up and tie your fingers!'* This was a reference to the cloth binding which had to be put on each finger, to protect the girls' hands from accidents with the sharp gutting knife, and from the curing salt. Any girl whose wounds were aggravated by the salt and failed to heal properly often had to give up the work with loss of income without compensation, and with the added expense of doctor's bills.

An experienced girl could gut and grade upwards of 60 herrings a minute, equivalent to 20,000 a day. Each barrel held betwen 700 and 1000 herrings, depending on size, and a packer could fill it with alternate layers of fish and salt, averaging three barrels an hour. Arranging the herring in the accepted pattern, with alternate layers at different angles, was a skilled but back-breaking job. Bottom and top layers were especially designed to show the herring at their best when the barrels were opened. At the bottom the black backs of the fish were displayed, while the silver bellies were displayed at the top. A very high standard of packing was not only demanded but achieved. The test of a well packed barrel was that the layers of herring remained in place even when the whole barrel was moved and placed on its side.

At the end of a season, £17 to £20 was considered a good average wage, a small sum perhaps in today's terms, but significant in those days when a cash-economy basis to living was gradually being introduced to remote mainland and island communities.

When the inflated bubble of the herring industry finally burst, it left large numbers of crofting communities balanced on a precarious knife-edge of survival. In some case, it was found that dependence on the annual appearance of the herring shoals round the coast had become so dominant that the land had suffered seriously from neglect and had reverted to a natural state, over-run by heather, bracken and rushes. Subsistence living reared its head, to continue the pattern of life which had been a feature of crofting.

Some men persevered and became full-time fishermen, regarding the croft as a useful means of income from sheep rearing and a source of potatoes as a staple food. But the erstwhile fertility of the crofts, growing corn, oats and turnips, slowly disappeared.

The annual returns of Scottish Fisheries show that the crofter-fisherman is still with us, but he is now a rare breed; only a very small percentage do more than fish spasmodically during the summer months, compared with over 10,000 nearly forty years ago, and three-quarters of Highland crofters a century ago.

The spectacular rise and equally spectacular decline of the fishing industry, and its devastating effect on the condition of crofting land, should have lessons for the crofter of today.

In those few decades, everyone concerned with the fishing industry got rich — except the Highlander. Once again, as in the years of the kelping industry, for example, others benefited from their skill and labour, and then left, and the Highlander perforce returned to a croft neglected and barren.

Surely there is a lesson for today in this. Nowadays the crofter only too often farms nothing but the tourist trade or something evanescent like the oil industry. They too will pass, as did the kelp and the herring. Only the land remains, always, and only the land and its proper husbandry can guarantee a certain future for those who hold it.

Looking at today's forest of Bed And Breakfast signs, flourishing amid a wasteland of neglected acres, one fears for the future.

SEVENTH SONS AND SEVENTH DAUGHTERS

Considering the peasant nature of Highland life before the middle of the 18th century, it might come as a surprise to discover that the Highland man, woman and child was generally a healthy specimen. *'Is e an oighreachd an t-slainte'* (health is the inheritance) is an old Gaelic proverb, one of many which suggest that concern for good health occupied the minds of the people. Setting aside sporadic problems of famine and reduced resistance to illness resulting from the scorched-earth policies of feuding clans, most people had little or no need for the services of a doctor or medical practitioner, qualified in a school or University in the south. There were, of course, those folk in most areas in the Highlands and Islands who had a special knowledge of herbs and potions for the cure of common ailments, and there were a few, members of some notable families, who were able to perform such delicate operations as the trepanning of skulls.

The First Statistical Account of Scotland (1790), made up of reports from the clergy in various parts of the Highlands, tells an interesting story.: *'We have commonly no sickness.'* (Dunoon): *'Very few diseases are known among the people.'* (Lochgoil): *No disease is peculiar to the parish from climate or any other cause.'* (Tarbert): *'Inhabitants all healthy.'* (Moy): *'We have no illness to speak of.'* (Loch Broom): *'The people have very few diseases.'* (Campbeltown). But occasionally disaster struck. The report in the Account on the district of Kilmaillie in Lochaber tells how a woman who had been harvesting in the south, brought home *'some low-country disorder'* which greatly upset the otherwise clean and healthy parish.

Donald MacLeod, in his book about the Highland Clearances, '*Gloomy Memories*', writes:

> 'I may mention that attendant on all previous and subsequent evictions, and especially this one, many severe diseases made their appearance, such as had hitherto been almost unknown among the Highland population, viz. typhus fever, consumption and pulmonary complaints in all their varieties, bloody flux, bowel complaints, eruptions, rheumatism, piles and maladies peculiar to women.'

It was quite clear in Donald MacLeod's mind that the aftermath of the Forty-Five and the era of the Clearances were major factors in the gradual deterioration in the standard of health of the folk in the Highlands and Islands.

References to physicians practising in the Highlands and Islands can be traced back to very early times, with the oldest Gaelic manuscript on medical matters bearing the date 1403. King Robert II of Scotland confirmed a charter, dated 1386, a grant of lands in the far north to Ferchard, the King's Leech. In 1444, there is mention of a '*leeche*' in the company of Sir Colin Campbell of Glenurchie, who was more than likely one of the noted medical family of the O'Conachers, who practised as hereditary doctors in Lorn. Another medical family were the Beatons, physicians to the Lords of the Isles. The family held lands in Islay, the seat of the old Lordship. After the Lordship was broken up, the Beatons scattered to enter the service of several Highland families: the MacLeans of Duart, the MacLeods of Skye and the MacDonalds of Sleat. Yet other branches of the family became physicians to the Frasers of Lovat, the Munros of Foulis and the old Earls of Sutherland. No doubt, with the clan system which obtained in the Highlands and Islands then, the services of the family physician were to some limited extent made available to the clansmen as well as to the chief's family.

Much of the medical knowledge of those physicians was written down in manuscript form, and a number of those writings survive today. Most of them date from the 14th to the 16th centuries and are full of quotations from eminent classical and contemporary medical authorities, from Hippocrates onwards. In 1305 a work called *'Lilium Medicinae'* was compiled by Bernard Gordon, a professor at the University of Montpelier. It became popular, not only on the continent of Europe, but also in Scotland, for shortly after its appearance, it was translated into Gaelic and several copies of this translation have survived. Tradition has it that one copy of it cost one of the Beatons of Skye no fewer than sixty milch cows, and so was highly prized. Indeed, so precious, was the book that when the doctor crossed an arm of the sea to visit a patient, the book was sent round by land. Another of the Beaton manuscripts, now in the British Museum, known as the *Regimen Sanitatis*, is thought to date from the 16th century.

This book begins by asserting that there were three aspects of the Regulation of Health: *Conseruatiuum* (guarding), *Preseruatiuum* (foreseeing), and *Reductiuum* (guiding backwards). The author explains that the first means the maintenance of a healthy state, the right of a healthy man; the second is the taking of the right measures by those in a declining state of health, and is therefore a duty; and the third aspect is the restoration of those who are ill, and is a necessity.

Many manuscripts deal with such ailments as fevers, diseases of the heart, anatomy, methods of diagnosis, and the properties of different plants. Inevitably, the manuscripts also contain references to the use of astronomy, astrology and metaphysics in medical practice. Some stress is laid on the use of blood-letting as a means to obtain cures in certain illnesses. The various books embody the highest knowledge available at the time, which speaks well of the professional knowledge of the Highland physician.

After the dispersal of those medical families, following the dissolution of the Lordship of the Isles, their descendents still kept their knowledge and practiced, as mentioned earlier, in various parts of the Highlands. Martin Martin mentions that when the Spanish Armada treasure ship *'Feorida'* was blown up in Tobermory harbour, Dr. MacBeth (Beaton), a well-known physician in Mull, was sitting on the upper deck of the ship, but survived the explosion. On Skye, Martin tells of the cures wrought by an '*illiterate empiric*', Neil Beaton, '*who of late is so well known in the isles and continent (Scottish mainland) for his great successes in curing several distempers, though he never appeared in the quality of a physician until he arrived at the age of forty years, and then also without the advantage of education.*' This Neil Beaton was credited with '*cutting a piece out of a woman's skull broader than a half-crown, and which restored her to perfect health.*'

In general, the poor of the Highlands and Islands had to make do with the administrations of local men and women with special knowledge of the curative properties of plants, usually members of families who had their knowledge and skill handed down from generation to generation. The seventh member of a family was supposed to be possessed of special gifts. This knowledge was very wide-ranging and is of long-standing. For instance, it is hardly an accident that the original Gaelic alphabet corresponds with the initials of the names of trees, perhaps an age-old indication of the worth of plants in the curing of ailments. The alphabet consisted of seventeen letters, to which was later added the letter 'H', though it is now used only for aspirative purposes.

The list is: *Ailm* (elm), *Beite* (birch), *Coll* (hazel), *Dur* (oak), *Eagh* (aspen), *Fearn* (alder), *Gath* (ivy), *Huath* (white-thorn), *Iogh* (yew), Luis (rowan), *Muin* (vine), *Nuin* (ash), *Oir* (spindle-tree), *Peith* (pine), *Ruis* (elder), *Suie* (willow), *Teine* (furze), *Ur* (health).

The plants which figured in the lore of the Gael served as a food, or else were used for their known healing properties; if the latter, they were administered in various forms: liquid to be drunk, or in an ointment base, severally, jointly, and often in concert with a ritual which contained some rhyme incorporating a supplication to a saint.

The knowledge of the herbal and medicinal uses of plants is of very long standing. It spans, more than likely, some millenia of tradition and factual lore, handed down in families who were regarded with respect in their communities and often given special positions and privileges in return for their services. The herbalists provided many basic concoctions, and housewives supplemented those by recipes based on plants. These latter included such dishes as *'cal dheanntag'* (nettle broth), and *'cal duilisg'* (dulse broth). In spring the first tender shoots of the common nettle were minced and boiled, sometimes with a little oatmeal. It was regarded as being one of the best spring diuretics, apart from being nourishing. *'Duilseag',* while it could be eaten raw (and indeed it warded off the hunger-pangs of many during the evictions of last century), was much more palatable and digestible when cooked. As a food and diuretic it was much prized by our maritime ancestors.

Other edible plants included various species of sorrel *(rumex: sealbhag),* still used in salad mixtures; the roots of the creeping cinquefoil or silverweed *(Potentilla anserina: brisgean, or barr brisgean).* The roots are tuberous and, eaten raw, have a slightly nutty taste. When roasted, however, they have a pleasant mealy flavour. The tuberous or everlasting bitter vetch (*Orobus tuberos: corr-meille)* has long branching roots, strung with nodulous lumps at frequent intervals. These, after drying, were chewed as wild liquorice. Far less deleterious (and much cheaper!) than common chewing gum, the taste continues in the mouth long after the last shred is chewed. The taste is both acid and sweet and so never palls. In times of scarcity the plant was used as a food:

> 'The Highlanders have a great esteem for the tubercules of the roots; they dry and chew them to give a better taste to their whisky. They also affirm that they are good against most diseases of the thorax, and that by the use of them they are enabled to repel hunger and thirst for a long time. In Breadalbane and Ross-shire, they sometimes bruise and steep them in water, and make an agreeable fermented liquor with them, called *Cairm.* They have a sweet taste like the roots of liquorice, and when boiled are well-flavoured and nutritive, and in times of scarcity have served as a substitute for bread.'

The range of wild plants used for the cure of ailments is very wide and embraces those which, consumed in quantity, would prove fatal, but when these are administered in small doses, in line with current medical practice, prove beneficial to the patient. No doubt many plants, poisonous in significant quantities, were introduced into community pharmacopoeias as the result of trial and error, but once their worth was proven, they became firm ingredients in remedies. With the advance of learning, the healing properties of plants came to be classified into a *materia medica* which, in the course of time, came to include remedies from the animal and mineral kingdoms. But the vegetable element in medicine persisted as the most important; indeed, not withstanding that many of the simple remedies used in past times are now discredited, they remain firmly embedded in the popular mind as being able to perform their traditional roles now as then.

The plants used for healing were divided into vulneraries, febrifuges, emetics, cathartics, irritants and tonics. In the first group, the vulneraries, are included cancer-wort (*Geranium rebertum: Lus-an-eallain),* used for skin afflictions: Kidney vetch (*anthillis Vulneraria: Meoir Mhuire)*, long held to be efficacious in the cure of cuts and bruises: Golden rod (*Solidago virgaurea: Fuinnseadh coille),* credited with the virtue of healing and joining broken bones: Wood sanicle (*Sanicula Europoea: Buine),* used in healing green wounds and ulcers: *Yarrow (Millefolium Europoeum: Earr-thalmhainn),* a potent styptic: Fig-wort (*Scrofularia nodosa: Lus-nan-cnap),* used in the cure of scrofula or King's Evil: and the highly regarded comfrey (*Symphylum officinalle: Meacan Dubh).* Old Culpeper, the famous sixteenth-century English herbalist, says of Comfrey:

> 'Yea, it is said to be so powerful to consolidate and knit together that, if they but boiled with dissevered pieces of flesh in a pit, it will join them together again.'

The second group, febrifuges, included the expellers of fever. Wood or dog violet (*Viola canina: Brog-na-cuthaig),* was used in a concoction boiled in whey to allay fevers: white helleborine (*Ipitatis litifolia: Ealobot geal)* cured colds in the head: blaeberry

(Vaccinium myrtillus: Lus-nam-broileag) soothed pains, and house-leek (*Sempervivum tectorum: Lus-nan-cluas)* was used, mixed with cream, to cure earache.

Counter irritants were used to reduce stubborn local pains. Among the best known were spearwort (a species of buttercup), which had to be used with caution on account of its violent action, and was attended with dangerous effects when administered internally. Martin Martin mentions that the best healer of the blisters raised by spearwort was a plaster of the sea plant *linnea-raich*, a species of *Confervae.* Groundsel *(Senecio vulgaris: Grunnasg)* was used for cataplasms to produce suppurations.

Many plants were used as emetics, some of the more popular being fir club moss (*Lycopodium selago: Garbhag-an-t-sleibhe),* which was effective, but had to be used with caution, and scurvy grass (Cochlearia officinalis: Am Maraich), which had a reputation as a valuable plant with medical properties of a corrective nature. The lesser meadow-rue (*Thalictrum minus: Ru Beag)* was a powerful cathartic.

Among the tonics and appetizers were lovage *(Ligusticum scoticum: Sunais),* much used in Martin Martin's time as a tonic and prompter of sluggish appetites, along with the gluttony-plant or dwarf cornel (*Cornus succosia: Lus-a-chraois),* and the dandelion (*Taraxacum leontodon: Am Bearnan Brighde),* which was, and perhaps still is, one of the most valuable ingredients in tonics and other medicines. Garden sage (*Salvia Officinalis: Saisde)* is remembered for its worth in healing by the Gaelic proverb: *'Carson a ghei-bheadh duine bas aig am bheil saisde fas no gharadh?'* (Why should a man die that has sage growing in his garden?) Self-heal or Heal-all (*Prunella vulgaris: Dubhan ceann -dubh)* was effective in removing all obstructions of the liver, spleen and kidneys: trefoil or bogbean (*Menyanthes trifoliata: Tri-bhileach)* was a potent tonic, often administered regularly in the form of a tisane.

If the knowledge of plants and their healing properties was well known, so also was that of plants which were poisonous. Hemlock, with its long-standing Greek reputation, had a number of names in

Gaelic suggesting some aspect of the plant. Night-weed was well-known for its large black berries and their somniferous qualities. More dangerous was henbane (*Hyascyamus niger: Gabhan),* sometimes known as *Cuthach-nan-cearc* ('that which sets the hens mad'). The juice of the petty splurge (*Euphorbia peplus: Gur-neimh)* was so caustic as to be used to destroy warts.

Perhaps it was inevitable, in the progression and development of the healing properties of plants, that medical knowledge came to be supplemented, and eventually complemented, with a belief in the magical properties of those same plants. Perhaps one or two failures, resulting from the wrong prescription for a particular illness, required some kind of guarantee that the applied remedy was to work successfully. It is common knowledge that the early Christian church humoured beliefs in the occult powers of plants, herbs and trees; but those beliefs, instead of being diminished by tolerance, grew in a body which assumed a new stature in folk lore and practice, by which time it had to be accepted with an even greater degree of tolerance. Yet, who is to say that the various chants, rhymes, runes and prayers said in accompaniment to the application of a remedy for a particular illness were not in themselves contributing to the success of the treatment? And the more so when those chants included supplication to saints, Biblical deities and even the Christ Himself?

However, beliefs went further, deep into the realm of protection against spiritual ills and ailments. For instance, Perforated St. John's wort (*Hypericum perforatum: caol aslachan Chaluim Chille)* was St. Columba's favourite flower; he reverenced it and carried it in his arms because it was dedicated to his favourite among the four evangelists, St. John. It was in great vogue in former times as a charm against witchcraft and enchantment. Another name for the plant, *Allas Muire* (meaning the image or semblance of the Virgin Mary), made it a rather sacred plant. The Gaelic name for Agrimony (*Agrimonia rupatoria: Mur Dhruid-hean)* reflected the association of the plant in the folk mind with the Druids, and its supposedly magical effects on spiritual troubles as well as diseases; it was used to heal a wide variety of ailments.

The rowan tree *(Pyrus aucuparia)* was known by various Gaelic names: *Luis* (drink) and *Luis-reog* (a charm which can be distilled into a spirit). Its presence beside homesteads all over the Highlands indicates, more then than now, the widespread belief that it was a charm against evil spirits. Lightfoot (1772) says the Highlanders believed '*that any part of the tree carried about with them proves a sovereign remedy against all the dire effects of enchantment or witchcraft.*'

The foxglove *(Digitalis purpurea)* had several Gaelic names associating the plant with the fairies, such as *Meuran sith* (Fairy thimble). The yarrow or milfoil *(Achillea millefolium: Earr-thalmhainn)* was an especially favoured plant with mystical properties. Young women cut it by moonlight with a black handled knife, for use in a ceremony which was designed to yield the name of their true loves. As with the rowan, the ash *(Fraxinus excelsior: Uinnseann)* was credited with magical properties and used against charms and enchantments. It was used, too, against the effects of serpent bites (spiritually as well as materially). Pennant (1772) says:

> 'In many parts of the Highlands, at the birth of a child, the nurse puts the end of a green stick of ash into the fire, and, while it is burning, receives into a spoon the sap or juice which oozes out at the other end, and administers this to the new-born babe.'

The combination of the known curative properties of plants and the suggestive curative elements inherent in supplication to saints and other propitiative beings became firmly established in the large body of folk medicine in the past. Administered as 'medical spells' for diseases of both man and beast, cures were, perhaps surprisingly, successful. Only on occasion were such ingredients used as bats' wings, dried spiders, and the like, so popular in the mind of the sceptic, not so much by genuine healers, who knew the tricks of their trade, but by quacks who imposed their will on their credulous clients, using tricks rather than traditional knowledge. Rather, the bulk of cures were effected by the use of materials, with complementary chants, which gave to the user the necesary degree of credibility. There were other methods, however, which used inanimate objects, such as beads and thread, rather than natural organic items. Relief for sprained ankles, dislocated joints and toothaches, was obtained by reciting a chant and passing a bead across the affected part, or a thread of common worsted, knotted at intervals, and as each knot passed over the fingers, a line of a charm was said. These 'cures' are still practised in the Highlands, though mainly with animals, which might be all the more significant because any possibility that auto-suggestion is involved is eliminated, and the cure must 'work' of its own accord, in its own way, and of its own volition.

Some cures are on record as authentic instances of the skill of the folk medicos. At the turn of the century a woman suffered from a rather prominent wen, which a neighbouring 'skilly woman' volunteered to remove for her. The remedy was simple: to walk over the moor to the skilly woman's house, a distance of about two miles. When she reached it, she was told to sit down, quite still, while the skilly woman got ready a needle and a knife. The needle was pointed at the wen, the knife made motions of cutting it off. The ceremony was accompanied by an incantation, in Gaelic, repeated seven times. After a few visits, always with the same ritual, the wen slowly disappeared, to the mystery and amazement of all who knew of the cure.

Blood-staunching required a gift known as *'casg fola'*, or blood-stopping. Any person with the gift could stop any bleeding at a moment's notice without medicines or appliance, but simply by the power of his word. It was, however, essential to know the name of the person to be cured, otherwise the charm would not work. There is on record the case of a lad in Wester Ross who cut himself so badly that his friends feared he would die from the loss of so much blood. They took him immediately to a local man, Iain Ban, who tried his charm, but it failed to work. It was only when another in the community, a woman, came forward with the suggestion that there were grave doubts as to the lad's paternity that the charm was administered again, with the lad's correct name uttered, when the bleeding stopped.

Another case involved a girl in the neighbourhood of Inverness who had a tooth extracted, which then bled for several hours. Treatment given by a qualified doctor brought only temporary relief. A friend then contacted a local farmer, known to possess the *casg fola,* who immediately went into a closet and muttered a charm. The time was 10.15 pm. Shortly he emerged and told the girl's friend to return, as the bleeding had stopped. In a fit of disbelief, the friend returned — to find that, indeed, the bleeding had been staunched, shortly after ten o'clock.

If there were no person with the gift in a district, natural means were used to staunch blood-flow, such as shepherd's purse, nettles, puff-balls, ribwort and spiders' webs, but yarrow was considered the best agent.

The first formal step towards the provision of health services in the Highlands and Islands was taken after the Poor Law (Scotland) Amendment Act of 1845, which placed upon parishes the duty of seeing that '*there shall be proper and sufficient arrangements made for dispensing and supplying Medicines to the sick Poor*', and of securing '*proper Medical Attendance for the Inmates of every.....Poorhouse.*' In addition, Parochial Boards were required '*to provide for Medicines, Medical Attendance, nutritious Diet, Cordials and Clothing for such Poor, in such Manner and to such Extent as may seem equitable and expedient.*'

These Statutory provisions were in fact of little relevance to conditions in the Highlands and Islands: it was not sufficiently realised that where money played little part in the local economy and the rateable values were invariably low, a gross disparity between the obligations and the resources available to meet them was unavoidable.

Nevertheless, such personal medical services as were provided in the Highlands were mainly attributable to this Act. But there was little real advance for many years until, by 1909, the Royal Commission on the Poor Law was told that there was a '*large amount of suffering in the more remote Highlands and Islands unrelieved by medical or nursing attendance.*'

One aspect of the conditions which the Highland crofting folk had to suffer is seen in the number of uncertificated deaths, which in turn gives an indication of the extent to which the level of medical attention during illness fell short of the desirable level. In 1881 the number of uncertificated deaths returned from the Scottish island parishes was 52% (compared with 17% in Scottish mainland rural districts and 10% for Scotland as a whole). By 1910 the respective figures had fallen to 25, 3 and 2 percent, indicating the improvement of services to the crofting population.

Today, despite the fact that many crofting communities are within reach of some medical service or facility, there are still problems which necessitate the calling for such special provisions as the Air Ambulance Service, yet one more aspect of the real face of crofting life.

A LIFE OF SONG

Whatever has been said of the crofting way of life, and whatever future historians may say about it, the musicologist will agree that its statutory creation in 1886 was a major factor in the survival to the present day of a large corpus of folk tradition and song which is almost without equal in Europe in its reflection of the history of a minority-language culture. And while the hard-headed crofter might look back at his ancestors' struggles with an iniquitous land-holding system in the Highlands and Islands, and think that his present day status was indeed sorely won, he might also agree that the struggles were worth it in terms of the continuity into the future of his language and culture — and even in the context of his identity as a member of an ethnic grouping in the British Isles, which has long been dominated by a majority language system.

The Gaelic-based culture of the Highlands and Islands is not alone in having, for instance, a large body of work songs, but it does tend to be unique in that those songs serve as an unbroken historical mirror in which the crofters' way of life is seen truly and honestly in all its aspects. Something of the same can be said for the poetry of many generations of Gaelic bards, whose work tells the story of those various events in Scotland's history which impinged on the life, and death, of the former clansmen.

A great deal, it might be argued, began with James Macpherson's '*Ossian*', a work which played such a significant role in the growth of European Romanticism, and which turned the eyes of Europe onto the culture of the Gaelic-speaking Highlanders, a culture which was fully a millenium old when Macpherson published, in 1762, his first book of purported ancient Gaelic heroic poetry. And even if Macpherson's work was later to fall out of respect, he at least sparked off an intense activity in the gathering of songs, poetry and traditions which is still going on today and has yielded such an amount of material that one cannot but admire the fact that while Highland folk starved, or were forced to seek new lives in countries across the seas, they still remembered what they had picked up at the hearth-stones of the storytellers and singers in the ceilidh-houses. On those memories grew the extant Gaelic culture in such overseas communities as one finds today in Nova Scotia.

At one time, each crofting community had one or more persons who were regarded as tradition-bearers. They were often members of a particular family in the township or district who had the gift of possessing prodigious memories, who could store a vast amount of material and were in their way the keepers of local tradition. This characteristic is still dominant where present day members of those families are now performing the roles of writers in their communities, using the media of stories, novels, poetry and song to record aspects of crofting life for those who are less articulate.

In older times, the *ceilidh* was more than just a gathering of neighbours to talk over local problems and discuss the import of whatever national news might have filtered through to the township via an itinerant packman or tinker. It was also the occasion when memories were revived, when the lineage of the folk of the township was re-stated, when the bards and singers were given the chance to freshen up old songs and poetry with yet another airing, and when the young were gradually made aware of their responsibilities to their families and to the community at large; and, not least, when the older folk of the township were taken into the heart of the communal spirit as manifested by the *ceilidh*. Thus, the *ceilidh* had a definite social function in the crofting township, and was not the occasion for pure entertainment as it tends to be today. The old-time *ceilidh* also served as the present-day community centre does, but was much less formal. The intense social character of such gatherings was fertile ground for the building up of character in the young folk so that they, too in their future time, could make the decisions which would allow them to survive the hardships of the crofting life.

The custom of the *ceilidh* gradually fell into disuse as the trappings of civilisation such as television became available in the crofting areas. But other influences also eroded the original function of the *ceilidh*: at Tolsta, in Lewis, the last *ceilidh-house* was burned down at its owner's wish as the emigrant ship, bearing him and his fellow-islanders, cruised up the coast and past the township. But it is interesting to note that the Western Isles Islands Council now includes in its plans for public housing in each township a *ceilidh-house* facility, where once again the custom of the social gathering can become part of village life.

In 1782, an English servant, accompanying his master on a visit to Skye, observed that the people of the district were so badly fed that they could not work well, but they also, to his mind, wasted much time telling:

> '.....idle tales and singing doggerel rhymes.....all kinds of labour is accompanied with singing: if it is rowing a boat the men sing; if it is reaping the women sing. I think that if they were in the deepest distress they would all join in a chorus.'

He was not far wrong, for the Gael has ever had the knack of singing out both his joy and distress through the medium of song. Apart from the songs used to accompany work (spinning, milking, churning, grinding flour and the like), there were songs used to record local love affairs, humorous and tragic incidents in the community, laments for those gone, and songs to celebrate momentous occasions in the district. Besides these, there were the great songs, sung to lyrics composed by the great Gaelic bards, such as Duncan Ban MacIntyre, celebrating the glorious Highland outdoors, and there were the political songs composed to fix in Highland history such events as the Jacobite Risings. Only some of many fine examples of these are to be found in the *Highland Songs of the Forty-Five*, edited by Dr. John Lorne Campbell of Canna, but they convey the very great intensity of feeling which such historical events engendered.

The spirit which produced the Jacobite movement also sparked off a remarkable creative industry which, even today, as in the poetry of George Campbell Hay, has not yet died. Contained in these songs are the heroism, the self-sacrifice, the valour and devotion to cause which has always marked off the Highlander from the rest of the inhabitants of the British Isles and which, later, was to make him such a useful tool in the creation of the British Empire, when for decades the Highland soldier was to be found in all corners of the earth fighting for British political and commercial interests — while at home his house and family were not even awarded stability of economy or security of tenure.

Songs are still being composed in the crofting townships, most having a relevance to the present day. An example of a modern bard, though thoroughly thirled to his past, is the late Murdo MacFarlane of Melbost in Lewis, whose poetic subjects ranged from the need to preserve the Gaelic language and culture to the atomic bomb and the nuclear age, from long, loving backward looks at the past to messages with a modern social content. And those songs are being sung and played by young Gaels who are becoming increasingly aware of the efforts and sacrifices of their forebears which fixed in perpetuity the crofting system and the way of life it supports, as almost a continuous statutory challenge to the regime of the landlord, a hunch to his back.

One body of working songs has survived to become the most important corpus of its kind in Western Europe: the waulking songs. *'Waulking'* was the process by which the woven cloth, straight from the weaver's loom, was thickened by a long and laborious process carried out by the women of the township. The waulking was an occasion where the folk of the township turned out in strength to make a *ceilidh,* but it also afforded the chance to freshen up a great store of traditional lore and re-commit to the memories of those who listened, the songs and stories recited at the event. There are some thousands of waulking songs in print and on record, representing only one facet of the great wealth of culture embedded in the memory of the older tradition-bearer.

The waulking was carried out with the aid of songs of a special type which allowed the waulking women to provide numerous choruses as they kneaded the cloth until it was thick enough to be fit for tailoring into a garment. The process is still carried out by hand in some areas, particularly in Harris, where the old-type, thick, heavy Harris Tweed is made. But it is fast becoming yet another aspect of the old days which have succumbed in the sands of time.

In a TV programme some time ago about life in the Western Isles, a Lewis bard, the late Murdo MacFarlane, said: 'Yet still we sing!' He was referring to the need for song in the life of the Gael today and the need for the native Gaelic speaker to retain much of the past fabric of Gaelic culture if the language is not to become a mere means of simple communication. Fortunately, agencies dedicated to the promotion of the Gaelic arts are now making significant inroads into the neglect of the past.

It is for this reason that the crofter is such an important element in the maintenance of the language and those aspects of its culture which are relevent to present day life and living. But even in those crofting areas where Gaelic is not spoken, such as in Caithness and the Northern Isles, the crofting way of life is an important sheet-anchor in the preservation of the spirit of independence. While it was not its primary purpose, the 1886 Crofters Act enabled the individual cultures of those communities to survive, to be instrumental in producing folk who are still proving to be the 'bank' of social values and ideas for the nation as a whole. And it is from those reserves of character that the will to survive against multinational commercial and national government interests is drawn. That indeed is a song worth the singing.

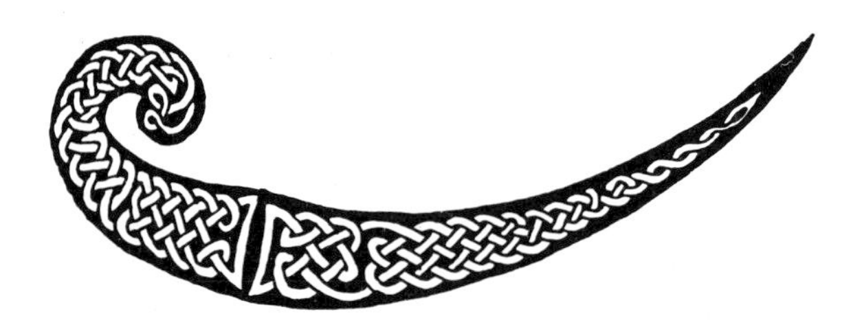

A POSTSCRIPT

It is commonly accepted that Acts of Parliament do not change things overnight, and the 1886 Crofters Act was no exception, though it did confer officially the name *'crofter'* on many who regarded that new status as some kind of blessing. The Act certainly did not offer the real change that was needed so desperately at that time: the enablement of crofters to work their neglected land back into the same degree of fertility that existed in previous centuries.

One might have thought that some seventy years of low-profile concern for change would have provided the necessary teeth to improve the crofters' lot by the time of the 1955 Crofters Act. But all that happened then was that improved grants for cropping and agricultural improvements were offered, which, in fact, only helped a minority of fertile crofts and large-acreage holdings (e.g. in Caithness and Orkney, as contrasted with the west coast of Scotland). Little account was taken of the socio-economic aspect of crofting. The model croft, if one interpretes legislation correctly, is one which aspires to the status of a small farm, which is an impossible dream for crofters on the east coast of Harris, for example. In 1976 yet more legislation was passed to give the crofter the right to buy his croft, an offer which only five per cent have taken up, the rest preferring to exist under the protection offered by the 1955 legislation.

It has been to the credit of the existing Crofters Commission that they have been concerned with encouraging the crofting community to become involved in commercial and industrial enterprises, so that the community can establish a stable economic base which, in turn, enhances the social aspects of crofting as both an activity and as a 'way of life'. The Commission once stated that the :

> ' ... willingness of crofters to work in industry combined with a desire to own their own homes and cultivate or stock a piece of land which they regard as their own, and which, for practical purposes, is their own, gives us an opportunity in the Highlands of working towards a new form of industrial society which will be healthier and more stable than any community which is completely urbanised. This is the true value of the small croft, and as new opportunities for employment are provided, it will increase rather than diminish.'

Some way along this route has been taken by some crofting communities which have established self-help organisations called Community Co-operatives. These have drawn on the strengths of their respective communities to create economic bases for the communal benefit. Not all have survived, but those which have have built up a pool of initiatives and entrepreneurial skills for use in the future.

The first Chairman of the Highlands and Islands Development Board is on record as saying: '*No matter what success is achieved in the Eastern or Central Highlands ... the Board will be judged by its ability to hold population in the true crofting areas.*' That the Board has failed to make any impact of significance on the crofting problem can be placed squarely on its attitude to land ownership in the Highlands. While accepting that the land itself cannot at present offer the crofters more than a fraction of the incomes they require, the land is, to quote from the Board's first Annual report, '*the Highlands' basic natural resource.*'

The future of crofting lies with the will and desire of the politician and the colour of his creed. Under one obvious colour, the politician has no desire to dismantle the system of land ownership in the Highlands, and wishes to leave the crofting community gradually to disappear as the whole weight of socio-economic disadvantage discourages people from staying in their communities. This would certainly be the future for those communities whose populations are already so distorted, being top-heavy with the aged.

Those politicians whose ideas bask under the red skies of a glowing sunset have had it in their grasp and power in the past to make good the neglect of decades, but little was done. Perhaps they feared that the capital sums released by public funds to compensate for land nationalisation might be used against them by the capitalist machine.

But other proposals have been put forward. Under a tartan flag, land would become the property of identified communities, and developed with public funds. On the basis that nothing suceeds like self-help, this idea seems to have some prospect for the future of crofting. It would certainly be of benefit to those crofters able to cope with the responsibility of developing their land resources to the full for the good of their own communities. And it would prevent some, at least, of the abuses which go on.

Since this book first appeared in 1984 there have been a number of significant developments on the crofting front. Perhaps the most exciting was the creation of the Scottish Crofters Union, a body dedicated to the furtherance of crofting interests and which has developed a hard-hitting political nose on land-use issues in the Highlands and Islands. The headline-grabbing ability of the Assynt Crofters to purchase their Island and to realise its economic potential was hailed as a pointer to the future of land use in the region.

The success, early in 1997, of the Isle of Eigg crofters' purchase of their island highlighted the evils of landlordism in the Highlands and Islands which, over some 170 years, saw the Eigg population fall from 400 to the current 65. The Eigg experience was particularly significant in that of the island's 7500 acres, only 1600 acres were actually under crofting tenure, unlike the Assynt buy-out which was mainly crofting land.

The future of crofting also lies in its ability to attract those of the younger generation willing to work hard to build up for themselves a stable foundation for their families. The New Croft Entrants Scheme, with over 200 on the waiting list with an average age of 28, is working successfully in its encouraging older crofters to release their crofts to suitable young aspirants.

The Crofter Forestry Act now on the Statute Book is helping to transform crofting. Trees were previously absent from croft land because any which grew above 5ft automatically became the laird's property. Tired land can now be transformed into fine young forestry by an imaginative and enterprising crofter, adding much needed value to crofting production. The fact that the Act was pushed through Parliament as a Private Member's Bill poses the question why it took so long for such an economic facility to provide the crofter with an addition to his long-term income. Since 1995 some one million new trees have been planted.

The offer of the Secretary of State for Scotland to release land under Scottish Office control to become Crofting Trusts is being looked at with caution but is yet another indication that, after more than a century, crofting as an environmentally-friendly form of land use has an ensured future. With care, attention, commitment. and not a little help from its friends, the land can win back its age-old basis as being a natural resource on which people can depend, to provide the warp against which other related activities can be woven to form the fabric of a rather unique way of life known as crofting.

APPENDIX

Statutory Powers For The Acquisition Of Land

Acquisition by the Secretary of State.

(One) Agriculture (Scotland) Act 1948.

(i) By Section 55 of this Act the Secretary of State may by agreement acquire (a) any land used for agriculture; (b) any other land which ought to be brought into use for agriculture; (c) additional land which is offered to him when he acquired under (a) and (b); and (d) any land which he could under the 1948 Act acquire compulsorily.

(ii) By Section 56 he can acquire compulsorily any land for the purposes of agricultural research or experiment or demonstrating agricultural methods.

(iii) By Section 57, the Secretary of State can compulsorily purchase land (a) Where full and efficient use of the land for agriculture is being prevented by reason of work not being carried out or fixed equipment not being provided, and it cannot reasonably be expected to be carried out or provided unless the Secretary of State exercises his powers; (b) where full and efficient use of land for agriculture will be prevented if existing fixed equipment thereon is not maintained and it cannot reasonably be expected to be maintained unless the Secretary of State exercises his powers; or (c) where, in exercise of Statutory powers, agricultural land has been or will be severed from other agricultural land for non-agricultural purposes and the full and efficient use of the severed land for agriculture cannot be achieved unless it is used in conjunction with other land.

(iv) The Secretary of State may also compulsorily acquire any additional land required to allow land acquired as above to be put to full and efficient use, and he must be prepared to secure the carrying out of what is required to be done.

(v) By Section 64, the Secretary of State can, by agreement or compulsorily, acquire land for land settlement.

(Two) Agriculture Act 1967

Section 45 of this Act extends the scope of Section 55 of the 1948 Act to include power to acquire land by agreement for the purposes of effecting amalgamations of agricultural land and re-shaping agricultural units.

Rural Development Board Powers

(1) Section 45 of the Agriculture Act 1967 provides that a Rural Development Board may be established for any area to apply the provisions contained in Part III of the Act for meeting the special problems of development for such rural areas as hills and uplands, and the special needs of such areas.
(2) Section 56 (2) of the 1967 Act provides that the Secretary of State may by order provide for the Highlands and Islands Development Board to exercise the powers and functions of a Rural Development Board in the Highlands and Islands or any part thereof.
(3) In terms of section 48 (1) of the 1967 Act a Rural Development Board can acquire by agreement land for purpose of effecting amalgamation and re-shaping agricultural units, *i.e.* the same powers as the Secretary of State has in terms of Section 29.

LUATH PRESS LIMITED

LUATH GUIDES TO SCOTLAND

Written by authors who invite you to share their intimate knowledge and love of the areas covered, these guides are not your traditional where-to-stay and what-to-eat books. They are companions in the rucksack or car seat, providing the discening visitor or resident with a blend of fiery opinion and moving description. Here you will find 'that curious pastiche of myths and legend and history that the Scots use to describe their heritage . . . A lively counterpoint to the more standard, detached guidebook . . . intriguing.'
The Washington Post

'Gentlemen,
We have just returned from a six week stay in Scotland. I am convinced that Tom Atkinson is the best guidebook author I have ever read, about any place, any time.'
Edward Taylor, Los Angeles

SOUTH WEST SCOTLAND: Tom Atkinson
ISBN 0 946487 04 9 pbk. £4.95

This descriptive guide to the magical country of Robert Burns covers Kyle, Carrick, Galloway, Dumfries-shire, Kirkcudbrightshire and Wigtownshire. Hills, unknown moors and unspoiled beaches grace a land steeped in history and legend and portrayed with affection and deep delight.
An essential book for the visitor who yearns to feel at home in this land of peace and grandeur.

THE LONELY LANDS: Tom Atkinson
ISBN 0 946487 10 3 pbk. £4.95

A guide to Inveraray, Glencoe, Loch Awe, Loch Lomond, Cowal, the Kyles of Bute and all of central Argyll written with insight,

sympathy and loving detail. Once Atkinson has taken you there, these lands can never feel lonely. 'I have sought to make the complex simple, the beautiful accessible and the strange familiar,' he writes, and indeed he brings to the land a knowledge and affection only accessible to someone with intimate knowledge of the area.
A must for travellers and natives who want to delve beneath the surface.

THE EMPTY LANDS: Tom Atkinson
ISBN 0 946487 13 8 pbk. £4.95

The Highlands of Scotland from Ullapool to Bettyhill and Bonar Bridge to John O'Groats are landscapes of myth and legend, 'empty of people, but of nothing else that brings delight to any tired soul,' writes Atkinson. This highly personal guide describes Highland history and landscape with love, compassion and above all sheer magic.
Essential reading for anyone who has dreamed of the Highlands.

ROADS TO THE ISLES: Tom Atkinson
ISBN 0 946487 01 4 pbk £4.95

Ardnamurchan, Morvern, Morar, Moidart and the west coast to Ullapool are included in this guide to the Far West and Far North of Scotland. An unspoiled land of mountains, lochs and silver sands is brought to the walker's toe-tips (and to the reader's fingertips) in this stark, serene and evocative account of town, country and legend.
For any visitor to this Highland wonderland, Queen Victoria's favourite place on earth.

HIGHWAYS AND BYWAYS IN MULL & IONA: Peter Macnab
ISBN 0 946487 16 2 pbk. £4.25

> 'The Isle of Mull is of Isles the fairest,
> Of ocean's gems 'tis the first and rarest.'

So a local poet described it a hundred years ago, and this recently revised guide to Mull and sacred Iona, the most accessible islands of the Inner Hebrides, takes the reader on a delightful tour of these rare ocean gems, travelling with a native whose unparalleled knowledge and deep feeling for the area unlock the byways of the islands in all their natural beauty.

THE SPEYSIDE HOLIDAY GUIDE: Ernest Cross
ISBN 0 946487 27 8 pbk. £4.95

Toothache in Tomintoul? Golf in Garmouth? Whatever your questions, Ernest Cross has the answers in this witty and knowledgeable guide to Speyside, one of Scotland's most popular holiday centres. A must for visitors and residents alike: there are still secrets to be discovered here!

WALK WITH LUATH

MOUNTAIN DAYS & BOTHY NIGHTS: Dave Brown and Ian Mitchell
ISBN 0 946487 15 4 pbk. £7.50

Acknowledged as a classic of mountain writing still in demand ten years after its first publication, this book takes you into the bothies, howffs and dosses on the Scottish hills. Fishgut Mac, Desperate Dan and Stumpy the Big Yin stalk hill and public house, evading gamekeepers and Royalty with a camaraderie which was the trademark of Scots hillwalking in the early days.
'The fun element comes through ... how innocent the social polemic seems in our nastier world of today ... the book for the rucksack this

year.' - **Hamish Brown, Scottish Mountaineering Club Journal**

'The doings, sayings, incongruities and idiosyncrasies of the denizens of the bothy underworld ... described in an easy philosophical style ... an authentic word picture of this part of the climbing scene in latter-day Scotland, which, like any good picture, will increase in charm over the years.' - **Iain Smart, Scottish Mountaineering Club Journal**

'The ideal book for nostalgic hillwalkers of the 60s, even just the armchair and public house variety ... humorous, entertaining, informative, written by two men with obvious expertise, knowledge and love of their subject.' - **Scots Independent**

'Fifty years have made no difference. Your crowd is the one I used to know ... [This] must be the only complete dossers' guide ever put together.' - Alistair Borthwick, author of the immortal ***Always a Little Further.***

THE JOY OF HILLWALKING: Ralph Storer
ISBN 0 946487 28 6 pbk. £ 6.95

Apart, perhaps, from the Joy of Sex, the Joy of Hillwalking brings more pleasure to more people than any other form of human activity.

'Alps, America, Scandinavia, you name it - Storer's been there, so why the hell shouldn't he bring all these various and varied places into his observations ... [He] even admits to losing his virginity after a day on the Aggy Ridge ... Well worth its place alongside Storer's earlier works.' - **TAC**

LUATH WALKING GUIDES

The highly respected and continually updated guides to the Cairngorms.
'Particularly good on local wildlife and how to see it' - **The Countryman**

WALKS IN THE CAIRNGORMS: Ernest Cross
ISBN 0 946487 09 X pbk. £3.95

This selection of walks celebrates the rare birds, animals, plants and geological wonders of a region often believed difficult to penetrate on foot. Nothing is difficult with this guide in your pocket, as Cross gives a choice for every walker, and includes valuable tips on mountain safety and weather advice.
Ideal for walkers of all ages and skiers waiting for snowier skies.

SHORT WALKS IN THE CAIRNGORMS: Ernest Cross
ISBN 0 946487 23 5 pbk. £3.95

Cross wrote this volume after overhearing a walker remark that there were no short walks for lazy ramblers in the Cairngorm region. Here is the answer: rambles through scenic woods with a welcoming pub at the end, birdwatching hints, glacier holes, or for the fit and ambitious, scrambles up hills to admire vistas of glorious scenery. Wildlife in the Cairngorms is unequalled elsewhere in Britain, and here it is brought to the binoculars of any walker who treads quietly and with respect.

BIOGRAPHY

ON THE TRAIL OF ROBERT SERVICE: Wallace Lockhart
ISBN 0 946487 24 3 pbk. £5.95

Known worldwide for his verses 'The Shooting of Dan McGrew' and 'The Cremation of Sam McGee', Service has woven his spell for Boy Scouts and learned professors alike. He chronicled the story of the Klondike Gold Rush, wandered the United States and Canada, Tahiti and Russia to become the bigger-than-life Bard of the Yukon. Whether you love or hate him, you can't ignore this cult figure. The book is a must for those who haven't yet met Robert Service.
'The story of a man who claimed that he wrote verse for those who wouldn't be seen dead reading poetry ... this enthralling biography will delight Service lovers in both the Old World and the New.' - **Scots Independent**

COME DUNGEONS DARK: John Taylor Caldwell
ISBN 0 946487 19 7 pbk. £6.95

Glasgow anarchist Guy Aldred died with 10p in his pocket in 1963 claiming there was better company in Barlinnie Prison than in the Corridors of Power. 'The Red Scourge' is remembered here by one who worked with him and spent 27 years as part of his turbulent household, sparring with Lenin, Sylvia Pankhurst and others as he struggled for freedom for his beloved fellow-man.
'The welcome and long-awaited biography of ... one of this country's most prolific radical propagandists ... Crank or visionary? ... whatever the verdict, the Glasgow anarchist has finally been given a fitting memorial.' - **The Scotsman**

BARE FEET & TACKETY BOOTS: Archie Cameron
ISBN 0 946487 17 0 pbk. £7.95

The island of Rhum before the First World War was the playground of its rich absentee landowner. A survivor of life a century gone tells his story. Factors and schoolmasters, midges and poaching, deer, ducks and MacBrayne's steamers: here social history and personal anecdote create a record of a way of life gone not long ago but already almost forgotten. This is the story the gentry couldn't tell.

'This book is an important piece of social history, for it gives an insight into how the other half lived in an era the likes of which will never be seen again' - **Forthright Magazine**

'The authentic breath of the pawky, country-wise estate employee.' - **The Observer**

'Well observed and detailed account of island life in the early years of this century' - **The Scots Magazine**

'A very good read with the capacity to make the reader chuckle. A very talented writer.' - **Stornoway Gazette**

SEVENS STEPS IN THE DARK: Bob Smith
ISBN 0 946487 21 9 pbk. £8.95

'The story of his 45 years working at the faces of seven of Scotland's mines ... full of dignity and humanity ... unrivalled comradeship ... a vivid picture of mining life with all its heartbreaks and laughs.' - **Scottish Miner**

Bob Smith went into the pit when he was fourteen years old to work with his father. They toiled in a low seam, just a few inches high, lying in the coal dust and mud, getting the coal out with pick and shovel. This is his story, but it is also the story of the last forty years of Scottish coalmining. A staunch Trades Unionist, one of those once described as "the enemy within", his life shows that in fact he has been dedicated utterly to the betterment of his fellow human beings.

HUMOUR/HISTORY

REVOLTING SCOTLAND: Jeff Fallow
ISBN 0 946487 25 1 pbk. £5.95

No Heiland Flings, tartan tams and kilty dolls in this witty and cutting cartoon history of bonnie Scotland frae the Ice Age tae Maggie Thatcher.
'An ideal gift for all Scottish teenagers.' - **Scots Independent**
'The quality of the drawing [is] surely inspired by Japanese cartoonist Keiji Nakazawa whose books powerfully encapsulated the horror of Hiroshima ... refreshing to see a strong new medium like this.' - **Chapman**

MUSIC AND DANCE

HIGHLAND BALLS & VILLAGE HALLS: Wallace Lockhart
ISBN 0 946487 12 X pbk. £6.95

Acknowledged as a classic in Scottish dancing circles throughout the world. Anecdotes, Scottish history, dress and dance steps are all included in this 'delightful little book, full of interest ... both a personal account and an understanding look at the making of traditions.' - **New Zealand Scottish Country Dances Magazine**
'A delightful survey of Scottish dancing and custom. Informative, concise and opinionated, it guides the reader accross the history and geography of country dance and ends by detailing the 12 dances every Scot should know - the most famous being the Eightsome Reel, "the greatest longest, rowdiest, most diabolically executed of all the Scottish country dances" .' **The Herald**
'A pot pourri of every facet of Scottish country dancing. It will bring back memories of petronella turns and poussettes and make you eager to take part in a Broun's reel or a dashing white sergeant!' - **Dundee Courier and Advertiser**

'An excellent an very readable insight into the traditions and customs of Scottish country dancing. The author takes us on a tour from his own early days jigging in the village hall to the characters and traditions that have made our own brand of dance popular throughout the world.'- **Sunday Post**

POETRY

THE JOLLY BEGGARS OR 'LOVE AND LIBERTY':
Robert Burns
ISBN 0 946487 02 2 hb. £8.00

Forgotten by the Bard himself, the rediscovery of this manuscript caused storms of acclaim at the turn of the 19th century. Yet it is hardly known today. It was set to music to form the only cantata ever written by Burns. **SIR WALTER SCOTT** wrote: 'Laid in the very lowest department of low life, the actors being a set of strolling vagrants ... extravagant glee and outrageous frolic ... not, perhaps, to be paralleled in the English language.' This edition is printed in Burns' own handwriting with an informative introduction by Tom Atkinson.
'The combination of facsimile, lively John Hampson graphics and provocative comment on the text makes for enjoyable reading.' - **The Scotsman**

POEMS TO BE READ ALOUD: selected and introduced by Tom Atkinson
ISBN 0 946487 00 6 pbk. £3.00

This personal collection of doggerel and verse ranging from the tear-jerking 'Green Eye of the Yellow God' to the rarely-printed bawdy 'Eskimo Nell' has a lively cult following. Much borrowed and rarely returned, this is a book for reading aloud in very good company, preferably after a dram or twa. You are guaranteed a warm welcome if you arrive at a gathering with this little volume in your pocket.
'The essence is the audience.' - **Tom Atkinson**

SELF-SUFFICIENCY

THE FAT OF THE LAND: John Seymour; illustrated by Sally Seymour
ISBN 0 9518381 0 5 pbk. £5.50

This was the seminal book by the Father of Self-Sufficiency in Britain. It is very much a how-to guide to the Good Life, filled to overflowing with advice from one who has done it all, and succeeded. It is written with the fervour and enthusiasm for which John Seymour is noted, and could be dangerous for your future: it would be very easy to decide to do what he and his wife and young children did and live literally on The Fat of the Land, almost entirely divorced from the stresses and strains of modern living.
'It is a favourite dream. I want to recommend this book to them all. It is a practical, down-to-earth account of one family's adventure in being self-supporting ... I don't know when I read a book that so gripped the imagination.' - **The Countryman**

Luath Press Limited

committed to publishing well written books worth reading

LUATH PRESS takes its name from Robert Burns, whose little collie Luath tripped up Jean Armour at a wedding and gave him the chance to speak to the woman who was to be his wife and the abiding love of his life. Burns called one of ***The Twa Dogs*** Luath after Cuchullin's hunting dog in Ossian's ***Fingal***. Luath Press grew up in the heart of Burns country, and now resides a few steps up the road from Burns' first lodgings in Edinburgh's Royal Mile.

Luath offers you distinctive writing with a hint of unexpected pleasures.

Luath Press Limited

543/2 Castlehill
The Royal Mile
Edinburgh EH1 2ND

Telephone (24 hours): 0131 225 4326
Fax: 0131 225 4324
email: gavin.macdougall@brainpool.co.uk

Please let us have your address (plus fax and email if applicable) if you would like us to keep you informed of future Luath publications.

Most UK bookshops either carry our books in stock or can order them for you. To order direct from us, please send a £sterling cheque, postal order, international money order or your credit card details (number, address of cardholder and expiry date) to us at the address above. Please add post and packing as follows: UK - £1.00 per delivery address; Europe airmail £2.00 per delivery address; overseas surface mail - £2.50 per delivery address; overseas airmail - £3.50 for the first book to each delivery address, plus £1.00 for each additional book by airmail to the same address. If your order is a gift, we will happily enclose your card or message at no extra charge.